Advances in Bioceramics and Biotechnologies

Advances in Bioceramics and Biotechnologies

Ceramic Transactions, Volume 218

*A Collection of Papers Presented at the
8th Pacific Rim Conference on Ceramic
and Glass Technology
May 31–June 5, 2009
Vancouver, British Columbia*

Edited by
Roger Narayan
Joanna McKittrick

Volume Editor
Mrityunjay Singh

A John Wiley & Sons, Inc., Publication

Published by John Wiley & Sons, Inc., Hoboken, New Jersey.
Published simultaneously in Canada.

For general information on our other products and services or for technical support, please contact our Customer Care Department within the United States at (800) 762-2974, outside the United States at (317) 572-3993 or fax (317) 572-4002.

Wiley also publishes its books in a variety of electronic formats. Some content that appears in print may not be available in electronic format. For information about Wiley products, visit our web site at www.wiley.com.

Library of Congress Cataloging-in-Publication Data is available.

ISBN 978-0-470-90548-7

Printed in the United States of America.

10 9 8 7 6 5 4 3 2 1

Contents

Al$_2$O$_3$ AND TiO$_2$

Preface

This volume is a collection of twenty-two cutting edge research papers from the symposia on Nano-Biotechnology and Ceramics in Biomedical Applications and Advances in Biomineralized Ceramics, Bioceramics, and Bioinspired Designs, which were presented at the 8th Pacific Rim Conference on Ceramic and Glass Technology (PACRIM-8) in Vancouver, British Columbia, Canada on May 31–June 5, 2009. PACRIM-8 is the eight in a series of international conferences on emerging ceramic technologies that began in 1993. PACRIM-8 was sponsored by The American Ceramic Society and was endorsed by The Australian Ceramic Society, The Ceramic Society of Japan, The Chinese Ceramic Society, and The Korean Ceramic Society. A total of 862 experts, practitioners, and end users from forty-one countries attended PACRIM-8, making it one of the most successful ceramic science and engineering events in recent years.

The symposia was focused on several key areas, including novel synthesis techniques, bioglasses and glass-ceramics, calcium phosphates for bone tissue applications, oxide ceramic implant applications. These papers cut across disciplines—ceramic science and technology, bioengineering and nanoscience—showing that a new, exciting field has emerged in the ceramics community.

We would like to thank Greg Geiger, Mark Mecklenborg, Marilyn Stoltz, and The American Ceramic Society for making this proceedings volume possible. We also give thanks to the authors, participants, and reviewers of this PACRIM proceedings issue. We hope that this issue becomes a useful resource in the area of bioceramics research that not only contributes to the overall advancement of this field but also signifies the growing roles of The American Ceramic Society and its partner ceramic societies in this rapidly developing field.

ROGER NARAYAN
JOANNA MCKITTRICK

Introduction

The 8th Pacific Rim Conference on Ceramic and Glass Technology (PACRIM 8), was the eighth in a series of international conferences that provided a forum for presentations and information exchange on the latest emerging ceramic and glass technologies. The conference series began in 1993 and has been organized in USA, Korea, Japan, China, and Canada. PACRIM 8 was held in Vancouver, British Columbia, Canada, May 31–June 5, 2009 and was organized and sponsored by The American Ceramic Society. Over the years, PACRIM conferences have established a strong reputation for the state-of-the-art presentations and information exchange on the latest emerging ceramic and glass technologies. They have facilitated global dialogue and discussion with leading world experts.

The technical program of PACRIM 8 covered wide ranging topics and identified global challenges and opportunities for various ceramic technologies. The goal of the program was also to generate important discussion on where the particular field is heading on a global scale. It provided a forum for knowledge sharing and to make new contacts with peers from different continents.

The program also consisted of meetings of the International Commission on Glass (ICG), and the Glass and Optical Materials and Basic Science divisions of The American Ceramic Society. In addition, the International Fulrath Symposium on the role of new ceramic technologies for sustainable society was also held. The technical program consisted of more than 900 presentations from 41 different countries. A selected group of peer reviewed papers have been compiled into seven volumes of The American Ceramic Society's Ceramic Transactions series (Volumes 212-218) as outlined below:

- **Innovative Processing and Manufacturing of Advanced Ceramics and Composites, Ceramic Transactions, Vol. 212,** Zuhair Munir, Tatsuki Ohji, and Yuji Hotta, Editors; Mrityunjay Singh, Volume Editor
  Topics in this volume include Synthesis and Processing by the Spark Plasma

Method; Novel, Green, and Strategic Processing; and Advanced Powder Processing

- **Advances in Polymer Derived Ceramics and Composites, Ceramic Transactions, Vol. 213,** Paolo Colombo and Rishi Raj, Editors; Mrityunjay Singh, Volume Editor
  This volume includes papers on polymer derived fibers, composites, functionally graded materials, coatings, nanowires, porous components, membranes, and more.

- **Nanostructured Materials and Systems, Ceramic Transactions, Vol. 214,** Sanjay Mathur and Hao Shen, Editors; Mrityunjay Singh, Volume Editor
  Includes papers on the latest developments related to synthesis, processing and manufacturing technologies of nanoscale materials and systems including one-dimensional nanostructures, nanoparticle-based composites, electrospinning of nanofibers, functional thin films, ceramic membranes, bioactive materials and self-assembled functional nanostructures and nanodevices.

- **Design, Development, and Applications of Engineering Ceramics and Composite Systems, Ceramic Transactions, Vol. 215,** Dileep Singh, Dongming Zhu, and Yanchun Zhou; Mrityunjay Singh, Volume Editor
  Includes papers on design, processing and application of a wide variety of materials ranging from SiC SiAlON, ZrO_2, fiber reinforced composites; thermal/environmental barrier coatings; functionally gradient materials; and geopolymers.

- **Advances in Multifunctional Materials and Systems, Ceramic Transactions, Vol. 216,** Jun Akedo, Hitoshi Ohsato, and Takeshi Shimada, Editors; Mrityunjay Singh, Volume Editor
  Topics dealing with advanced electroceramics including multilayer capacitors; ferroelectric memory devices; ferrite circulators and isolators; varistors; piezoelectrics; and microwave dielectrics are included.

- **Ceramics for Environmental and Energy Systems, Ceramic Transactions, Vol. 217,** Aldo Boccaccini, James Marra, Fatih Dogan, Hua-Tay Lin, and Toshiya Watanabe, Editors; Mrityunjay Singh, Volume Editor
  This volume includes selected papers from four symposia: Glasses and Ceramics for Nuclear and Hazardous Waste Treatment; Solid Oxide Fuel Cells and Hydrogen Technology; Ceramics for Electric Energy Generation, Storage, and Distribution; and Photocatalytic Materials.

- **Advances in Bioceramics and Biotechnologies, Ceramic Transactions, Vol. 218;** Roger Narayan and Joanna McKittrick, Editors; Mrityunjay Singh, Volume Editor
  Includes selected papers from two cutting edge symposia: Nano-Biotechnology and Ceramics in Biomedical Applications and Advances in Biomineralized Ceramics, Bioceramics, and Bioinspiried Designs.

I would like to express my sincere thanks to Greg Geiger, Technical Content Manager of The American Ceramic Society for his hard work and tireless efforts in the publication of this series. I would also like to thank all the contributors, editors, and reviewers for their efforts.

MRITYUNJAY SINGH
Volume Editor and Chairman, PACRIM-8
Ohio Aerospace Institute
Cleveland, OH (USA)

Synthesis Techniques

A ROLE FOR ELECTROCHEMICAL SYNTHESIS IN BIOCERAMIC COMPOSITE MATERIALS

M. Ramesh Kumar[a], Erika F. Merschrod S.*[,a], Kristin M. Poduska*[,a,b]
[a]Department of Chemistry and [b]Department of Physics and Physical Oceanography,
Memorial University of Newfoundland, St. John's, Newfoundland and Labrador, Canada

ABSTRACT

We present an overview of the motivations, desirable characteristics, and synthesis methods for ceramic materials with potential orthopaedic applications. In this context, we summarize our recent investigations of the electrochemically assisted synthesis of calcium phosphate and collagen-calcium phosphate composite materials. Electrochemical methods offer the opportunity to use pH differences and electric fields, rather than thermal energy, to control mineral crystallization and protein aggregation. Electrodeposition generally requires the use of conducting substrates, but we have also used electrochemistry to develop suspended, substrate-free collagen membranes, which form the basis for our composite materials. Through a combination of complementary characterization techniques, including Raman scattering spectroscopy (for phase characterization in poorly crystallized minerals) and force spectroscopy (for protein aggregation information), we address the details of composition, structure and formation in mineral, collagenous and composite materials.

INTRODUCTION

Ceramic materials play an important part in biologically relevant systems.[1] Since bone is based on apatitic calcium-phosphate mineral phases, medical orthopaedics have benefitted greatly from studies of the strength and bioactivity of ceramic and ceramic composite materials.[2] These studies bridge medical and materials sciences because the implant materials are developed in the laboratory (*ex situ*) but their true function must be evaluated in the body (*in situ*).[3] The focus of this paper is an overview of some *ex situ*, lab-developed materials that are promising candidates to improve medical implant coatings in bone-replacement contexts, followed by highlights of our recent work using electrochemical methods to make controlled bioceramic composite materials.

BONE AS A CERAMIC MATERIAL

Natural bone is a complex and hierarchically structured composite material based on hydroxyapatite (HAp), $Ca_{10}(PO_4)_6(OH)_2$ (HAp), with Mg and carbonate substitutions. The distinctive combination of strength and light weight that is ascribed to bone comes not from the inherent strength of the ceramic, but rather from the way the composite material is structured.[4,5] Structure and strength are intimately related to the presence of protein, primarily Type I collagen, and other organic components.[6]

There have been many studies that have focused on the macroscopic structure and strength of bone.[5-7] These are complemented by other studies of the nucleation and crystallization of bone-like apatitic minerals.[8-12] However, there is still much to be investigated (and understood) in terms of how the mesoscale structure and properties of bone relate to bone regeneration, in both natural and artificial environments.[13-16]

TARGET PROPERTIES FOR EFFECTIVE BONE IMPLANTS AND COATINGS

Implant performance in the body is largely a matter of durability and functionality in the end product.[15] A few of the most widely addressed characteristics of implant coatings are described below.

Table I. Selected calcium phosphate phases

Mineral name	Abbreviation	Composition
hydroxyapatite	HAp	$Ca_{10}(PO_4)_6(OH)_2$
dahllite		$Ca_5(PO_4, CO_3)_3(OH)$
brushite		$CaHPO_4 \cdot 2H_2O$
monetite		$CaHPO_4$
octacalcium phosphate	OCP	$Ca_8(HPO_4)_2(PO_4)_4 \cdot 5H_2O$
tricalcium phosphate	TCP	$Ca_3(PO_4)_2$
amorphous calcium phosphate	ACP	$Ca_3(PO_4)_2 \cdot xH_2O$

Crystallinity and phase

It is interesting that, although there are many different phases of calcium phosphate (Table I), only apatitic phases are found in bone.[4, 8] Studies show that new bone, whether in infants or in bone produced during healing in adults, is poorly crystallized.[17] As bone develops, a compositional change develops as the bone crystallizes. Since many implant coatings are based on amorphous calcium phosphate, it is not yet clear whether specific calcium phosphate phases present advantages (or disadvantages) in terms of implant coating performance.[18]

Confirming the phase composition of calcium phosphates is challenging because of the many possible crystalline and amorphous phases that are known to exist. As a result, it is often advantageous to use crystalline structure determination techniques (X-ray diffraction) in tandem with studies of more localized order (Raman scattering spectroscopy).[11] In general, HAp does not crystallize well at low temperatures, whether prepared synthetically or produced in the body.[17] Other phases, such as brushite, produces very crystalline electrodeposits with well-defined Bragg diffraction peaks.[18]

Solubility

Since bone growth in the body occurs in an aqueous environment, the solubility of the implant coating can impact the chemical environment in which bone regrowth occurs. Solubility of the coating can be influenced by the crystallinity and crystal structure of the coating, as well as the presence of other incorporants in a composite coating.[8, 19]

Roughness and texture

Natural bone is a hierarchical material comprised of organic and mineral components. As the bone growth occurs, the interactions between cells and the implant can be affected by the topographic cues on the coating. There are a number of examples in the literature that indicate that cell growth can be helped or hindered by morphological features on a surface.[14, 16, 20]

Strength

Many ceramic coatings on implants are designed to produce a biologically familiar surface to an implant designed for strength, such as a metal bone pin or joint. Therefore, the strength does not necessarily need to come from the ceramic material itself. However, there are instances in which the strength is designed into the ceramic composite materials.[21] The mechanical adherence and wear properties of a coating are also important to prevent degradation (which could reduce the mechanical integrity of the implant, as in the case of joint replacements) or disintegration (which can trigger biological processes that interfere with bone growth).[2]

Bioactivity

Ultimately, the role of an implant coating is to increase the likelihood that the implant is seamlessly incorporated into the body.[22, 23] Clinical studies have shown that calcium phosphate-based coatings dramatically accelerate the body's incorporation of metal pins.[2] In contrast, reduced bioactivity can be important on areas such as ball and socket joints, wherein the tissue buildup impedes the use of the implant.

Another approach to improving the bioactivity of an implant coating is to embed specific incorporants in a composite to tailor the interaction between the implant and the body.[1, 3, 14] For example, since natural bone contains both mineral and proteinaceous components, it may be desirable to incorporate proteins such as collagen into calcium phosphate based implant coatings; this has been addressed with theoretical[24] and experimental[5, 7] studies. Incorporating antibiotics in the coatings can reduce the risk of infection[22, 25] when coating solubility can be regulated to sustain the benefit of the antibiotic.

OVERVIEW OF SYNTHESIS METHODS FOR BONE IMPLANTS AND COATINGS

Developing an implant coating that displays all of the attributes described above is a challenge. The synthesis of these coatings often involves multiple steps, each of which can be grouped into one of three broad categories. For the most part, these syntheses produce stoichiometric calcium phosphates (HAp, brushite, OCP, or TCP), although there are also numerous of studies devoted to looking at Mg-substituted or carbonated apatites. There are a number of recent reviews focused on bioceramic synthesis, both stoichiometric and substituted.[1, 8, 10, 26]

Thermal synthesis

For control of crystallinity and particle size, thermal synthesis methods are well developed for many ceramic materials. Calcium phosphates have a complex phase diagram, in part because there are so many different crystalline phases that exist.[8] Hydrothermal, spray pyrolysis, and conventional high-temperature treatments have all been used to make bone-like implant coatings.[26] These offer the possibility of thermodynamic selectivity of different calcium phosphate phases, as well as the option of controlling crystallite size and uniformity. Annealing studies have been performed under either inert or reactive atmospHeres in order to control the composition. An alternative approach is to chemically dissolve or mechanically grind natural bone into small particles, to then be coated onto a substrate and heat treated for optimal adherence.[27]

Other thermal studies have used high temperature processes to control morphology as well as composition. Micrometer-scale bubbles have been used to induce bone-like porosity into ceramic materials, and sacrificial incorporants that can be thermally removed by combustion have also been used to create porous structures.[12]

Since many organic or pharmaceutical additives are not stable at the high temperatures used in these thermal synthesis techniques, such incorporants are typically added in a separate, low-temperature step as a top layer to the implant coating.

Solution-phase synthesis

Given that bones can grow in the body at relatively low temperatures, some methods are meant to mimic the body's growth environment through the use of biological molecules used by cells to stimulate and organize the mineralization process.[12, 28] Others are designed to use chemistry that is very different from what goes on in the body, such as the use of alkaline treatments to control phase selectivity.[26] Because alkaline treatments can cause denaturation of protein incorporants, multi-step coating preparation processes are often employed.

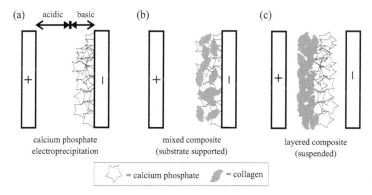

Figure 1: Schematic depictions of electrochemical precipitation of (a) calcium phosphate, (b) a mixed calcium phosphate-collagen composite, and (c) a layered calcium phosphate-composite. It is possible to produce both mixed and layer composites as either substrate-supported or suspended materials. In all three schemes, the cathode is denoted with a minus (–) and the anode is labeled with a plus (+).

Electrochemically assisted synthesis

An extension of the chemical methods for calcium phosphate crystallization involves the use of an applied electric potential to trigger an electrochemical crystallization reaction. This kind of hydrolysis-induced precipitation reaction is well-known,[10,29–31] but its application to the controlled synthesis of metal oxide and other ceramic materials is far less studied than its use in metal electroplating. Electrodeposition offers the option of using applied potential to control the energetics of the reaction without need for high temperatures. Therefore, it provides the possibility of controlling both kinetics and thermodynamics in a single, low-temperature reaction process.

Electrodeposition presents both advantages and limitations. The electrochemical precipitation process occurs only in the immediate vicinity of the substrate (working electrode) surface, so this means that, in general, electrically conductive substrates are required. For metal bone pins and similar stainless-steel or titanium alloy implants, this is not a problem. In fact, electrodeposition offers the capability of coating complex electrode shapes due to the conformal nature of the electrodeposit. Because the energetics of the precipitation reaction are influenced by the relative reduction potentials of different metal ions (dependent on electrolyte pH and ionic strength of the electrolyte), not all metals can be easily incorporated into an electrodeposit, and compositional control of alloys can be complicated. Nevertheless, there are examples from our group[11,32] and others[31,33] of phase-selective materials synthesis procedures using electrochemical methods.

ELECTROCHEMICALLY PREPARED BIOCERAMICS

We have developed electrochemical methods to produce tailored bioceramic coatings. This includes phase-selective calcium phosphate minerals on stainless steel substrates,[11] as well as collagen-supported calcium phosphate composite materials.[34,35]

Calcium phosphate coatings on metal substrates

With many crystalline phases of calcium phosphate minerals known to exist, we sought a method to selectively form specific calcium phosphate phases. Electrochemical methods have been used by our group[32] and others to produce different metal oxide phases by controlling applied deposition potential, based on a mechanism where the applied potential produces an oxidation state change for ions in solution, which then triggers the electrodeposition process.[29,30] In the case of calcium phosphates, however, we use electrochemical hydrolysis to create a spatially localized pH change in the electrolyte (shown schematically in Figure 1a) where the alkaline pH conditions trigger a calcium phosphate precipitation process. This mechanism has been used to synthesize other metal oxide materials such as ZnO[31,36] and Cu_2O,[33] as well calcium phosphate coatings.[10,37–39] One of our contributions to this field was demonstrating that electrosynthesis conditions (more specifically, the buffering capacity of the electrolyte) can be used to selectively produce two different calcium phosphate phases in a one-step deposition process, without the need for alkaline or high-temperature treatments.[11]

The key to this control is evident upon closer analysis of the chemistry involved in the deposition procedure. The electrochemically assisted precipitation of HAp requires an alkaline environment. We utilized a nitrate reduction reaction (–0.23 V *vs.* a saturated calomel electrode), as others have, to accelerate the electrogeneration of base in acidic bulk electrolyte.[38] This ensured that the increased OH^- concentration was spatially confined to the working electrode surface, preventing precipitation from the bulk electrolyte.

$$NO_3^- + H_2O + 2\,e^- \longrightarrow NO_2^- + 2\,OH^- \qquad (1)$$

$$H_2PO_4^- + OH^- \longrightarrow HPO_4^{2-} + H_2O \qquad (2)$$

$$10\,Ca^{2+} + 6\,HPO_4^{2-} + 8\,OH^- \longrightarrow Ca_{10}(PO_4)_6(OH)_2 + 6\,H_2O \qquad (3)$$

When there is insufficient OH^- present, brushite forms at the expense of HAp. Our work demonstrates that a simple way to control the amount of OH^- present is to adjust the buffering capacity of the electrolyte through changing hydrogen phosphate concentration.[11] The equilibria of different hydrogen phosphate species in solution is very complex, but the role they play in the electroprecipitation of hydroxyapatite is consistent with phosphate supersaturation.[38,40,41] Interestingly, calcium phosphates prove to be much more resistant to epitaxial growth compared to calcium carbonates[42] or other metal oxide materials.[31,33]

Collagen-supported calcium phosphate coatings

We have also investigated how to tailor the build-up of a protein-mineral composite materials using electrochemical methods. One possibility is to use a layering approach, while another is to make a mixed material; these options are shown schematically in Figure 1b,c. A mixed calcium phosphate-collagen composite does not form readily using a simple electrolyte composed of collagen monomers dispersed in a standard[11,39] nitrate-based calcium phosphate electrolyte. Such deposits show very little evidence of collagen incorporation according to Raman spectroscopic data. Alternatively, we used electrochemical methods to create a thin layer (~100–400 nm thick) of monomeric collagen directly on stainless steel. Our investigations suggest that monomers, rather than the larger and more structured protofibrils or fibrils, preferentially adsorb to the electrode due to their small size and hence faster migration.[43] Calcium phosphate can then be electrodeposited on top of this layer. Attempting to put a thin collagen layer on top of electrodeposited calcium phosphate yields very inconsistent results, likely due to the potential drop across the poorly conducting calcium phosphate layer.

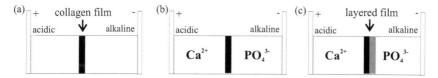

Figure 2: Schematic depiction of electrosynthesized layered calcium phosphate-collagen composite. A suspended collagen membrane is formed at the mid-point between parallel plate electrodes (a), after which calcium and phosphate ions are added to the cathode and anode sides of the membrane, respectively. When the applied electric field is sustained, calcium phosphate precipitates on the alkaline (anode) side of the collagen membrane. The centimeter-sized membrane can then be removed and dried on a substrate of choice.

A different approach to composite formation that lends itself to both the layered and mixed material approaches involves creating a collagen scaffold that does not require a supporting substrate. We accomplished this by using isoelectric focusing in an electrochemical cell with parallel plate electrodes and a collagen-rich electrolyte.[34] Application of a large electric potential (relative to the stability of water) causes acid and base generation at the counter and working electrodes, respectively. A pH difference will change the net charge on a collagen monomer due to protonation/deprotonation, and this in turn affects the electrostatic interaction between the monomer and the applied electric field. As a result, there is a net migration of collagen away from both electrodes toward the region of the electrolyte whose pH matches the isoelectric point of collagen ($\sim$pH 6). When the concentration of collagen in the solution is high enough, this migration culminates in a collagen scaffold. These macroscopic scaffolds can be removed from the electrochemical cell and dried on a material of choice. We expect that it is possible to incorporate other kinds of particles in the membrane as it is forming, either through similar charge-based interactions with the applied electric field, or by particles being trapped in the migrating collagen.[10] It is interesting to note that applying isoelectric focusing to other biomaterially relevant proteins such as fibronectin or albumin does not yield a robust, integral membrane.

To incorporate calcium phosphate into the scaffold, Ca^{2+} and PO_4^{3-} ions can be added to the electrolyte after membrane formation; subsequently, a calcium phosphate layer forms on the alkaline side, as shown schematically in Figure 2. We investigated composite formation using a range of calcium- and phosphate-based salts. A representative set of Raman scattering data (Figure 3, collected with 532 nm excitation from a confocal Jobin Yvon Horiba LabRAM) shows that HAp is the preferred phase for composites prepared with a $CaCl_2$ precuror, regardless of our choice of phosphate salt (KH_2PO_4, K_2HPO_4, or $NH_4H_2PO_4$). Similar results were obtained using $Ca(NO_3)_2$.

An advantage of substrate-free composite materials is that they could be applied to cell scaffolding studies[14] where metal implants are not required. In this context, we have developed a method tune and measure the stiffness of the scaffold, prior to forming the calcium phosphate layer,[35] which augments the range of elastic responses possible for the final composite material. Stiffness is one parameter that has been used to direct the response of cells, including their growth, regeneration, and differentiation.[20]

We used indentation measurements with an atomic force microscope (MFP-3D, Asylum Research) to assess stiffness changes in the collagen membranes as a function of ion content in the electrolytes used to aggregate the collagen. Force curves were collected at ambient temperature and humidity using silicon probes (MikroMasch NSC35, spring constant $\sim$17 nN/nm). The curvature of the extension and retraction curves was accounted for entirely by probe shape and did not show evidence of coupling with the harder substrate. We then employed simplified, but widely utilized, purely elastic models[35]

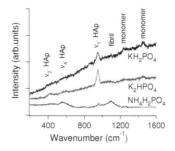

Figure 3: Raman spectroscopic data for layered calcium-phosphate composites prepared from one of three different phosphate salts (KH_2PO_4, K_2HPO_4, or $NH_4H_2PO_4$) and 10 mM $CaCl_2$. Peaks indicative of hydroxyapatite (labeled HAp) and collagen (labeled fibril or monomer) are evident in all samples. Spectra are offset along the intensity axis for clarity, and peak assignments are based on previous studies.[44,45]

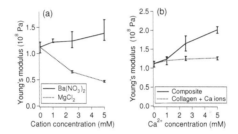

Figure 4: (a) Young's moduli values can change dramatically with the addition of different concentrations of ions such as Mg^{2+} or Ba^{2+} during membrane formation. (b) Young's moduli of calcium phosphate-collagen composites are higher than those of membranes prepared in the presence of calcium ions alone. In both plots, the lines connecting data points serve as guides to the eye.

that neglect viscoelastic effects in order to extract Young's modulus information.[46–49]

Our investigations show that adding different ions to the electrolyte have dramatically different effects on the stiffness of the resulting collagen scaffold, suggesting that ion content in the electrolyte has a strong effect on the collagen assembly process. For example, adding different doubly charged cations to the electrolyte can make either stiffer or softer films, as shown in Figure 4a. Earlier work from our group has shown that Ca^{2+} increases stiffness, K^+ decreases stiffness and Na^+ has a minimal effect.[35] Membrane stiffness becomes even more pronounced after the calcium phosphate precipitation step (Figure 4b).

CONCLUSION
Electrochemical methods offer a very flexible toolbox with which we can fine tune the structural, chemical, and mechanical properties of calcium phosphate–collagen composite coatings. We are now beginning to assess clinically related target properties of our electrochemically synthesized materials, including cell adhesion.

FOOTNOTES
* Corresponding authors. E-mail: kris@mun.ca, erika@mun.ca

ACKNOWLEDGEMENTS
Thanks to Memorial's Department of Biochemistry for the use of their Raman spectroscopy facilities. We thank NSERC (Canada) and the Canada Foundation for Innovation for funding support.

REFERENCES

[1] S. M. Best, A. E. Porter, E. S. Thian, and J. Huang, Bioceramics: Past, present and for the future, *J. Eur. Ceram. Soc.*, **28**, 1319–1327 (2008).

[2] H. R. Piehler, The future of medicine: Biomaterials, *MRS Bull.*, **25**, 67–70 (2000).

[3] D. F. Williams, On the mechanisms of biocompatibility, *Biomater.*, **29**, 2941–2953 (2008).

[4] S. Weiner and H. D. Wagner, The material bone: structure-mechanical function relations, *Annu. Rev. Mater. Sci.*, **28**, 271–298 (1998).

[5] L. C. Palmer, C. J. Newcomb, S. R. Kaltz, E. D. Spoerke, and S. I. Stupp, Biomimetic systems for hydroxyapatite mineralization inspired by bone and enamel, *Chem. Rev.*, **108**, 4754–4783 (2008).

[6] M. J. Olszta, X. Cheng, S. S. Jee, R. Kumar, Y.-Y. Kim, M. J. Kaufman, E. P. Douglas, and L. B. Gower, Bone structure and formation: A new perspective, *Mater. Sci. Engr. R*, **58**, 77–116 (2007).

[7] F.-Z. Cui, Y. Li, and J. Ge, Self-assembly of mineralized collagen composites, *Mater. Sci. Eng. R*, **57**, 1–27 (2007).

[8] L. Wang and G. H. Nancollas, Calcium orthophosphates: Crystallization and dissolution, *Chem. Rev.*, **108**, 4628–4669 (2008).

[9] S. R. Qiu and C. A. Orme, Dynamics of biomineral formation at the near-molecular level, *Chem. Rev.*, **108**, 4784–4822 (2008).

[10] I. Zhitomirsky, Cathodic electrodeposition of ceramic and organoceramic materials. Fundamental aspects, *Adv. Colloid Interface Sci.*, **97**, 279–317 (2002).

[11] S. D. Huelin, H. R. Baker, E. F. Merschrod S., and K. M. Poduska, Phase-selective electroprecipitation of calcium phosphate thin films at physiological temperatures, *Cryst. Growth Des.*, **6**, 2634–2636 (2006).

[12] F. C. Meldrum and H. Cölfen, Controlling mineral morphologies and structures in biological and synthetic systems, *Chem. Rev.*, **108**, 4332–4432 (2008).

[13] J. E. Davies, Bone bonding at natural and biomaterial surfaces, *Biomater.*, **28**, 5058–5067 (2007).

[14] M. M. Stevens and J. H. George, Exploring and engineering the cell surface interface, *Science*, **310**, 1135–1138 (2005).

[15] B. D. Boyan, C. H. Lohmann, D. D. Dean, V. L. Sylvia, D. L. Cochran, and Z. Schwartz, Mechanisms involved in osteoblast response to implant surface morphology, *Annu. Rev. Mater. Res.*, **31**, 357–371 (2001).

[16] F. Barrère, T. Mahmood, K. de Groot, and C. van Blitterswijk, Advanced biomaterials for skeletal tissue regeneration: instructive and smart functions, *Mater. Sci. Eng. R*, **59**, 38–71 (2008).

[17] C. Meneghini, M. C. Dalconi, S. Nuzzo, S. Mobilio, and R. H. Wenk, Rietveld refinement on X-ray diffraction patterns of bioapatite in human fetal bones, *Biophys. J.*, **84**, 2021–2029 (2003).

[18] L. B. Gower, Biomimetic model systems for investigating the amorphous precursor pathway and its role in biomineralization, *Chem. Rev.*, **108**, 4551–4627 (2008).

[19] M. T. Fulmer, I. C. Ison, C. R. Hankermayer, B. R. Constantz, and J. Ross, Measurements of the solubilities and dissolution rates of several hydroxyapatites, *Biomater.*, **23**, 751–755 (2002).

[20] D. E. Discher, P. Janmey, and Y.-L. Wang, Tissue cells feel and respond to the stiffness of their substrate, *Science*, **310**, 1139–1143 (2005).

[21] K. Balani, R. Anderson, T. Laha, M. Andara, J. Tercero, E. Crumpler, and A. Agarwal, Plasma-sprayed carbon nanotube reinforced hydroxyapatite coatings and their interaction with human osteoblasts in vitro, *Biomater.*, **28**, 618–624 (2007).

[22] S. E. Sakiyama-Elbert and J. A. Hubbell, Functional biomaterials: design of novel biomaterials, *Annu. Rev. Mater. Res.*, **31**, 183–201 (2001).

[23] R. Zapanta LeGeros, Calcium phosphate-based osteoinductive materials, *Chem. Rev.*, **108**, 4742–4753 (2008).

[24] J.-W. Shen, T. Wu, Q. Wang, and H.-H. Pan, Molecular simulation of protein adsorption and desorption on hydroxyapatite surfaces, *Biomater.*, **29**, 513–532 (2008).

[25] M. Vallet-Regí, F. Balas, M. Colilla, and M. Manzano, Bioceramics and pharmaceuticals: A remarkable synergy, *Solid State Sci.*, **9**, 768–776 (2007).

[26] Y. Gao and K. Koumoto, Bioinspired ceramic thin film processing: present status and future perspectives, *Cryst. Growth Des.*, **5**, 1983–2017 (2005).

[27] K. Haberko, M. M. Bućko, J. Brzezińska-Miecznik, M. Haberko, W. Mozgawa, T. Panz, A. Pyda, and J. Zarębski, Natural hydroxyapatite – its behaviour during heat treatment, *J. Eur. Ceram. Soc.*, **26**, 537–542 (2006).

[28] A. George and A. Veis, Phosphorylated proteins and control over apatite nucleation, crystal growth, and inhibition, *Chem. Rev.*, **108**, 4670–4693 (2008).

[29] J. A. Switzer, Electrochemical synthesis of ceramic films and powders, *Am. Ceram. Soc. Bull.*, **66**, 1521–1524 (1987).

[30] G. H. A. Therese and P. V. Kamath, Electrochemical synthesis of metal oxides and hydroxidess, *Chem. Mater.*, **12**, 1195–1204 (2000).

[31] D. Lincot, Electrodeposition of semiconductors, *Thin Solid Films*, **487**, 40–48 (2005).

[32] S. Chatman, A. J. G. Noel, and K. M. Poduska, Tuning magnetic hysteresis of electrodeposited Fe_3O_4, *Journal of Applied Physics*, **98**, 113 902–1–6 (2005).

[33] J. A. Switzer, R. Liu, E. W. Bohannan, and F. Ernst, Epitaxial electrodeposition of a crystalline metal oxide onto single-crystalline silicon, *Journal of Physical Chemistry B*, **106**, 12 369–12 372 (2002).

[34] H. R. Baker, E. F. Merschrod S., and K. M. Poduska, Electrochemically controlled growth and positioning of suspended collagen membranes, *Langmuir*, **24**, 2970–2972 (2008).

[35] M. R. Kumar, E. F. Merschrod S., and K. M. Poduska, Correlating mechanical properties with aggregation processes in electrochemically fabricated collagen membranes, *Biomacromolecules* in press (2009).

[36] T. Ren, H. R. Baker, and K. M. Poduska, Optical absorption edge shifts in electrodeposited ZnO thin films, *Thin Solid Films*, **515**, 7976–7983 (2007).

[37] Y. Fan, K. Duan, and R. Wang, A composite coating by electrolysis-induced collagen self-assembly and calcium phosphate mineralization, *Biomaterials*, **26**, 1623–1632 (2005).

[38] M. Shirkhanzadeh, Direct formation of nanophase hydroxyapatite on cathodically polarized electrodes, *J. Mater. Sci.: Mater. Med.*, **9**, 67–72 (1998).

[39] G. H. A. Therese, P. V. Kamath, and G. N. Subbanna, Novel electrosynthetic route to calcium phosphate coatings, *J. Mater. Chem.*, **8**, 405–408 (1998).

[40] D. B. Haddow, M. S. Thompson, S. R. Berry, and J. T. Czernuszka, Compositional and structural control in bone regenerative coatings, *J. Mater. Sci. – Mater. Med.*, **10**, 219–222 (1999).

[41] X. Lu, Z. Zhao, and Y. Leng, Calcium phosphate crystal growth under controlled atmosphere in electrochemical deposition, *J. Crystal Growth*, **284**, 506–516 (2005).

[42] E. A. Kulp and J. A. Switzer, Electrochemical biomineralization – the deposition of calcite with chiral morphologies, *J. Am. Chem. Soc.*, **263**, 15 120–15 121 (2007).

[43] S. D. Huelin, H. R. Baker, K. M. Poduska, and E. F. Merschrod S., Aggregation and adsorption of type I collagen near an electrified interface, *Macromolecules*, **40**, 8440–8444 (2007).

[44] P. N. de Aza, F. Guitián, and C. Santos, Vibrational properties of calcium phosphate compounds. 2. Comparison between hydroxyapatite and β-tricalcium phosphate, *Chem. Mater.*, **9**, 916–922 (1997).

[45] M. Wisniewski, A. Sionkowska, H. Kaczmarek, S. Lazare, V. Tokarev, and C. Belin, Spectroscopic study of a krf excimer laser treated surface of the thin collagen films, *J. Photochem. Photobio. A: Chem.*, **188**, 192–199 (2007).

[46] B. B. Akhremitchev and G. C. Walker, Finite sample thickness effects on elasticity determination using atomic force microscopy, *Langmuir*, **15**, 5630–5634 (1999).

[47] J. L. Hay and P. J. Wolff, Small correction required when applying the Hertzian contact model to instrumented indentation data, *J. Mater. Res*, **16**, 1280–1286 (2001).

[48] G. M. Pharr, W. C. Oliver, and F. R. Brotzen, On the generality of the relationship among contact stiffness, contact area, and elastic modulus during indentation, *J. Mater. Res.*, **7**, 613–617 (1992).

[49] W. C. Oliver and G. M. Pharr, Measurement of hardness and elastic modulus by instrumented indentation: Advances in understanding and refinements to methodology, *J. Mater. Res.*, **19**, 3–20 (2004).

ELECTRO THERMALLY POLARIZED SINTERED BULK-HAP AND HAP-COATED Ti FOR BIOMEDICAL APPLICATIONS

Subhadip Bodhak, Susmita Bose and Amit Bandyopadhyay*
W. M. Keck Biomedical Materials Research Laboratory, School of Mechanical and Materials Engineering, Washington State University, Pullman, WA 99164, USA.
*Corresponding author: amitband@wsu.edu

ABSTRACT
    Electro-thermally polarized sintered bulk HAp compacts can store electrostatic surface charge which has shown to improve the osseointegration of implants by accelerating the early stage mineralization and tissue in-growth processes. In this research we have investigated the efficacy of polarization treatments for HAp coated Ti samples and compared their polarizability as well as in vitro mineralization behavior with that of polarized bulk sintered HAp compacts. A uniform 20 ± 1.38 μm thick HAp coating was made on Ti substrate with HAp slurry using a spin coater and the bulk HAp compacts were prepared by sintering at $1200°C$ for 2h. Both the samples were polarized via the application of an external d.c. field of 2 kV/cm at $400°C$ for 1h. Thermally stimulated depolarization current (TSDC) analysis revealed that less dense HAp coating samples stored lower surface charge (1.69 μC/cm^2) than bulk sintered HAp compacts (4.28 μC/cm^2) under the same polarization condition. It has been observed that increasing the surface potential led to decreasing hydrophobicity of HAp surface irrespective of sample nature. In vitro mineralization study revealed that accelerated apatite crystal growth was occurred on the negatively charged surface, whereas apatite growth was inhibited on the positively charged surface for both bulk and coated HAp samples.

INTRODUCTION
    Hydroxyapatite (HAp) ceramics have been well researched over the past three decades for repair and reconstruction of bone defects, and are commonly used for coatings and defect fillers [1, 2]. Because of its compositional and crystal structure similarity to that of natural bone, synthetic HAp ceramics have the ability to bond directly to the living bone and can conduct the formation of new bone on their surfaces [3]. However, as the synthetic HAp has a lower osteogenic or chemical bone-bonding capability compare to the living bone tissue, the major problem concerning the success and lifetime of these HAp based implants in physiological condition is their low degree of osseointegration and/or tissue integration properties at the implant-tissue interface which delay the bone reconstruction process as well as the healing time [4]. Therefore, the current trend in HAp based artificial bone substitutes design and production is to produce implants with tailored surface properties to accelerate the bone reconstruction process as well as the healing time. The manipulation of implant surfaces is becoming necessary since the surface properties dominate the early stage in vivo implant-tissue interactions [5]. Recent advances in biomaterial research have revealed that electro thermally polarized bulk sintered HAp can produce significant biological response in both in vitro [6-8] and in vivo [9] conditions. It has been demonstrated that the adhesion and adsorption properties of HAp surface can be greatly manipulated by means of electrical polarization treatment. According to these studies, a large polarization charge between 0.08 μC/cm^2 to 1.2 mC/cm^2 can be stored within sintered HAp compact by applying a d.c. electric field at elevated temperature [10-12]. In our recent studies, we have shown that the negatively polarized HAp surface favored early stage mineralization while the positively charged surface inhibited the apatite nucleation under in vitro condition [13]. Furthermore, we have noticed that enhanced bone like apatite layer deposition on negatively charged surface promoted the human fetal osteoblasts (hFOB) cell attachment by favoring the anchorage between cellular microextensions i.e., lamelllipodia or filopodia to the substrate which led to increased surface occupancy of osteoblast cells [13]. This clearly indicates that polarized hydroxyapatite with tailored

surface properties can provide a new basis for the development of a new generation of implants that can promote faster healing. However, to date very few reports are available on polarized HAp coated implants [14, 15]. Therefore, in our current research, we aim to investigate the efficacy of polarization treatments for HAp coated Ti implant. The present investigation is also focused on understanding the influence of surface charge, wettability, and charge polarity on bone like apatite layer deposition on polarized HAp coating surface in SBF since early stage mineralization is crucial for implant stability and success of the musculoskeletal implants. Finally, a comparison of polarizability and bioactivity properties of electrically polarized sintered bulk HAp and coated HAp -Ti samples has also been made to determine the role of induced electrostatic surface charge in influencing the *in vitro* mineralization behavior of HAp ceramics.

MATERIALS AND METHODS

Hydroxyapatite powders were synthesized in-house by sol gel route. Calcium nitrate tetrahydrate [Ca(NO$_3$)$_2$, 4H$_2$O] and diammonium hydrogen phosphate [(NH$_4$)$_2$HPO$_4$] were used as precursor materials [13]. A Ca: P ratio of 1.67 was maintained to obtain stoichiometric HAp phase. After calcinations at 1000°C for 1 h, to prepare bulk sintered HAp compacts, HAp powders were subsequently pressed in a hydraulic uniaxial cold press to form discs (12 X 1mm) and finally the discs were sintered at 1200°C for 2 h in air. The sintered discs were smoothly polished using diamond paste upto 1μm surface finish and then ultrasonically cleaned before use. For HAp coating sample preparation commercially pure Ti (99.6 at.%) (Alfa Asear, MA, USA) discs of 12 mm in diameter and 0.5 mm in thickness were used as the substrate. A 40 wt% solid loaded HAp slurry was prepared by dispersing sol gel synthesized HAp powder in HAp sol. The slurry was then used in HAp coating preparation using a spin coater (Laurell Tech. Corp., PA, USA). Slurry based thick film preparation was done to gain greater control over coating thickness and homogeneity [16]. After depositing 4 layers thick HAp coatings, samples were heat treated in a muffle furnace (Thermo Fisher, MA, USA) at 700°C for 15 min. Surface microstructures and coating thickness were observed under a field emission scanning electron microscope (FESEM, FEI, OR, USA).

The prepared samples were then electro thermally polarized in air, using platinum electrodes, with Keithley 6487 picoammeter (Keithley Instruments Inc., Ohio, USA) and ceramic strip heater (OMEGA Engineering Inc., Connecticut, USA). During polarization, both bulk sintered HAp compacts and HAp coated samples were heated up to a polarization temperature (Tp) at 400°C and then a d.c. electrical field (Ep) of 2.0 kV/cm was applied for 1h at Tp and maintained until the sample slowly cooled to room temperature. In order to estimate the stored static charge due to polarization operation, thermally stimulated depolarization current (TSDC) technique was used [17]. In this method polarized samples were heated from room temperature to 550°C, with a heating rate of 5°C/min and the thermally stimulated release depolarization current was measured using Keithley 6487 picoammeter. The stored electrical charge was calculated from the TSDC spectra using the following equation [10].

$$Qp = \frac{1}{\beta} \int J(T)dT \qquad (1)$$

Where, Qp denotes the stored charge, β and J(T) are the heating rate and the current density, respectively.

To measure the wettability of samples, the static contact angles of sessile drops of D.I. water on sample surfaces were measured using a face contact angle set-up equipped with a camera (VCA Optima, AST Products Inc., MA, USA). Surface energies of bulk sintered HAp compacts were calculated from the contact angle data as obtained for two polar liquids, formamide and glycerol, and one apolar liquid, diiodomethane.

The influence of stored surface charge and charge polarity on *in vitro* mineralization behavior of both polarized HAp coated Ti and bulk sintered HAp compacts were evaluated by immersion in simulated body fluids (SBF) which has a similar ionic concentration to human blood plasma. All samples were immersed in the glass vial containing SBF solutions and were kept under static conditions inside a biological thermostat at 37°C. After immersion for a maximum of 5 days samples were observed under a field emission scanning electron microscope (FESEM). All experiments were performed in triplicate, by running three independent glass vials simultaneously. From here on, the negatively charge surface will be termed as 'N' poled surface and positively charged surface as 'P' poled surface.

RESULTS AND DISCUSSIONS

Fig. 1 presents the representative SEM microstructural images for both sintered bulk HAp compact and HAp coating surfaces. An average grain size between 9 and 10 μm was measured for sintered HAp compact as shown in Fig. 1a. It can be recalled that the starting HAp powder particle size was between 1 and 2 μm. The grain size was increased during solid state sintering process. Figure 1b shows 4 layers thick HAp coating surface microstructure. An average coating thickness of 20±1.38 μm was determined from the crosssectional SEM image. Although the coating was free from surface cracking and delamination, the porous morphology of the coating surface can be clearly noticed. Compared to the coating microstructure, the relative density of the bulk sintered HAp compacts was measured approximately 91% of theoretical density. It can be mentioned that both the porosity and the grain size in sintered body are found very important for electrical properties of these materials [18, 19].

Fig. 2 shows the TSDC curves as obtained for sintered bulk HAp compacts and HAp coating samples polarized at 400°C under application of a d.c. electrical field of 2.0 kV/cm for 1h. From TSDC results we confirmed the success of the polarization treatment as the depolarization currents were released from HAp compact or coating surfaces at elevated temperature indicating HAp polarizability [17]. However, in spite of identical polarization condition, the marked difference in TSDC spectra can be clearly noticed between sintered bulk HAp and coating samples. For the bulk HAp sample (Fig. 2a) a broad and strong TSDC curve was observed which increased at 403°C, reached maximum points at 510°C corresponding to a maximum current density (J) peak of 4.37 nA/cm^2. But for the HAp coating sample, a narrow and weak TSDC curve (Fig. 2b) was obtained which started increasing at 435°C, reached maximum points at 493°C showing a maximum current density (J) peak of only 1.19 nA/cm^2, and then gradually decreased. This clearly suggests the low electrostatic charge storage ability of HAp coating compared to bulk sintered HAp compact samples. Table 1 presents the calculated stored charges from the TSDC spectra of polarized samples. Bulk sintered HAp compacts exhibited much higher stored charge density (Qp) of 4.28μC/cm^2 than that of HAp coating samples (Qp ~ 1.69μC/cm^2). This difference in charge storage ability can be well explained from the differences in sample microstructure and physical properties. It can be recalled that polarization of HAp has been attributed to ionic conduction mechanism, where proton (H$^+$) was considered as the principle charge carrier at low temperature (T$_p$ ~ upto 400°C) [10-13]. Under application of an electric field at elevated temperature, proton was migrated along the columnar (OH$^-$) channels of hexagonal unit cell of HAp and caused a charge separation within HAp crystal structure. This electro thermal polarization phenomenon subsequently resulted in accumulation and/or storage of opposite polarity electrostatic charge on two surfaces of HAp compacts or coatings. However, the polarizability of HAp depends on several factors such as polarization temperature (T$_p$), polarizing electric filed (E$_v$), grain size, presence of porosity, phase purity of the material etc. [10-13, 20]. In this work, the difference in charge storage ability of bulk sintered HAp compact and HAp coating can be attributed to the low volume of HAp for coated sample which is only 20±1.38 μm thick compared to 1 mm thick sintered HAp compacts. In addition, less dense porous microstructures were also responsible for lower polarizability in HAp coated samples. It has been assumed that higher porosity restricted the proton transport and thus reduced the charge separation. Whereas the large grain microstructure of dense sintered polycrystalline

HAp led to reductions in grain boundary and therefore eventually reduced the resistivity of the material and increased polarizability.

Table 1 presents the measured contact angles with water for both polarized sintered HAp compacts and HAp coating samples. For comparison the contact angle and surface energy data for unpoled samples were also given in Table 1. Experimental results showed that HAp surface wettability and surface energy can be tailored by inducing surface charge without introducing any volumetric effects in the material. It has been demonstrated that increasing the surface potential led to decreasing of hydrophobicity of HAp surface. During polarization the changes in surface electronic state were assumed to modify the surface energy as well as surface wettability of polarized HAp ceramics. For both 'N' and 'P' poled sintered HAp compacts a significant decrease in contact angle (upto ~ 18°) was observed. However, the difference among poled and unpoled sample wettability was not very distinct for HAp coating surfaces because of porous morphology and presence of surface irregularities. The influence of surface charge on surface energy of polarized HAp compacts can also be clearly seen from Table 1. Surface energies as high as 49.47 mJ/m^2 and 48.67 ± 0.87 mJ/m^2 were observed on 'P' poled and 'N' poled HAp compacts compared to 35.20 mJ/m^2 for unpoled HAp compact surfaces. It can be recalled that an implant with higher wettability and increased surface energy enabled easy deposition and nucleation of apatite crystal in both *in vitro* and *in vivo* conditions [21].

To understand the influence of surface charge, wettability and charge polarity on bone like apatite deposition, we have investigated *in vitro* SBF mineralization behavior of both polarized bulk sintered HAp and HAp coated Ti samples. Fig. 3 presents SEM surface morphology of both 'N' poled and 'P' poled samples after 5 days of immersion in SBF. Interestingly, accelerated growth of bone like apatite layer was clearly observed on highly wettable 'N' poled sintered HAp compacts (Fig. 3a) and HAp coating surfaces (Fig. 3c). This observation is in good agreement with previous reports [6, 7, 13]. It is believed that on 'N' poled HAp surface, because of negative surface charge Ca^{2+} ions preferentially absorbed and subsequently formed a supersaturated ionic solution with other anionic groups present (i.e., HPO$_4^{2-}$, HCO$_3^-$, OH$^-$). This in turn helped in selective deposition of bone like apatite layers on 'N' poled HAp surfaces [13]. However, the higher charge storage on polarized bulk sintered HAp surface (Fig. 3a) favored a thick bone like apatite layer deposition compared to 'N' poled HAp coated Ti surface (Fig. 3c). A high magnification SEM image shown in Fig. 3a indicates a thin plate like HAp crystallite deposition. In contrast, on 'P' poled surfaces apatite nucleation was markedly decelerated and salt like crystal deposition was predominantly observed (Fig. 3b and d). It is believed that anions such as Cl$^-$ are selectively attracted to positively charged surface and reacted with Na$^+$ ions present in SBF solution to deposit NaCl salts on the surface [13]. These findings have provided evidence that induced electrostatic surface charge and its polarity can efficiently control *in vitro* mineralization behavior of polarized sintered HAp compacts and HAp coated Ti samples.

CONCLUSIONS

In this study we have compared the polarizability and charge storage capability of sintered bulk HAp compacts and HAp coated Ti samples. Furthermore, the possible influence of electrostatic surface charge and its polarity on *in vitro* mineralization behavior of electro thermally polarized bulk sintered and coated HAp samples was also investigated. The TSDC measurement indicated that less dense HAp coating stored lower charge than sintered HAp compacts. Wettability and surface energy calculations revealed that increasing the surface potential led to increasing of wettability and surface energy of HAp surfaces. *In vitro* SBF immmersion study revealed that 'N' poled surface accelerated the mineralization process while 'P' poled surface inhibited apatite nucleation during initial days of immersion for both bulk HAp and HAp coated samples. Therefore, early stage mineralization on negatively charged HAp surfaces can potentially induce accelerated healing of the injured bone tissue after implantation.

ACKNOWLEDGEMENTS
Authors like to acknowledge financial support from the Office of Naval Research under the grant no N00014-01-05-0583.

REFERENCES
[1]M. Jarcho, Calcium phosphate ceramics as hard tissue prosthetics, *Clin. Orthop. Relat. Res.*, **157**, 259-78 (1981).
[2]A. Bandyopadhyay, S. Bernard, W. Xue, and S. Bose, Calcium Phosphate-Based Resorbable Ceramics: Influence of MgO, ZnO, and SiO2 Dopants, *J. Am. Ceram. Soc.*, **89 [9]**, 2675-88 (2006).
[3]L. L. Hench, An introduction to bioceramics (Advanced Series in Ceramics), *World Scentific, Singapore: Wilsoln*, **1**, 1-24 (1993).
[4]A. J. Salgado, O. P. Coutinho, and R. L. Reis, Bone Tissue Engineering: State of the art and future trends, *Macromol Biosci.*, **4**, 743-65 (2004).
[5]K. Yamashita, and S. Nakamura, Concept and development of vector ceramics for bio-interface engineering, *J. Ceram. Soc. Jap.*, **113 [1]**, 1-9 (2005).
[6]K. Yamashita, N. Oikawa, and T. Umegaki, Acceleration and deceleration of bone-like crystal growth on ceramic hydroxyapatite by electric poling, *Chem. Mater.*, **8**, 2697-700 (1996).
[7]P. Calvert, and S. Mann, The negative side of crystal growth, *Nature*, **386**, 127-128 (1997).
[8]M. Ohgaki, T. Kizuki, M. Katsura, and K. Yamashita, Manipulation of selective cell adhesion and growth by surface charges of electrically polarized hydroxyapatite, *J. Biomed. Mater. Res. A*, **3 [57]**, 366-73 (2001).
[9]T. Kobayashi, S. Nakamura, and K. Yamashita, Enhanced osteobonding by negative surface charges of electrically polarized hydroxyapatite, *J. Biomed. Mater. Res.*, **57**, 477—84 (2001).
[10]S. Nakamura, H. Takeda, and K. Yamashita, Proton transport polarization and depolarization of hydroxyapatite ceramics, *J. Appl. Phys.*, **10 [89]**, 5386-92 (2001).
[11]M. Ueshina, S. Nakamura, and K. Yamashita, Huge, millicoulomb charge storage in ceramic hydroxyapatite by bimodal electric polarization, *Adv. Mater.*, **14 [8]**, 591-95.
[12]J. P. Gittings, C. R. Bowen, I. G. Turner, F. R. Baxter, and J. B. Chaudhuri, Polarisation behaviour of calcium phosphate based ceramics. *Mater. Sc. Forum*, **587-588**, 91-95 (2008).
[13]S. Bodhak, S. Bose, and A. Bandyopadhyay, Role of surface charge and wettability on early stage mineralization and bone cell-materials interactions of polarized hydroxyapatite, *Acta Biomater.*, (2009); In Press: doi: 10.1016/j.actbio.2009.02.023.
[14]R. Kato, S. Nakamura, K. Katayama, K. Yamashita, Electrical polarization of plasma-spray–hydroxyapatite coatings for improvement of osteoconduction of implants, *J. Biomed. Mater. Res. A*, **74A [4]**, 652-58 (2005).
[15]T. Kobayashi, S. Itoh, S. Nakamura, M. Nakamura, K. Shinomiya, and K. Yamashita, Enhanced bone bonding of hydroxyapatite-coated titanium implants by electrical polarization, *J. Biomed. Mater. Res. Part A*, **82A [1]**, 145-51 (2007).
[16]O. Huang, A. Bandyopadhyay, and S. Bose, Influence of processing parameters on PZT thick films, *Mater. Sc. Eng. B*, **116** 19-24 (2005).
[17]C. Bucci, R. Fieschi, and G. Guidi, Ionic Thermocurrents in Dielectrics, *Phys. Rev.*, **148**, 816-18 (1966).
[18]M. Oztas, Influence of grain size on electrical and optical properties of InP films, *Chin. Phys. Lett.*, **25 [11]**, 4090-92 (2008).
[19]M. S. Dash, J. Bera, and S. Ghosh, Effect of porosity on electrical properties of undoped and lanthanum doped BaTi, *Int. Conf. Sol. Dielec.*, 211-284 (2007).
[20]J. P. Gittings, C. R. Bowen, A. C. E. Dent, I. G. Turner, F. R. Baxter, and J. B. Chaudhuri, Electrical Characterization of hydroxyapatite-based bioceramics, *Acta Biomater.*, **5 [2]**, 743-54 (2009).

[21]K. Das, S. Bose, and A. Bandyopadhyay, Surface modifications and cell–materials interactions with anodized Ti, *Acta Biomater.*, *3*, 573-85 (2007).

Table I. Calculated stored charge, maximum current density, measured contact angle with water and surface energy values for polarized and unpoled bulk sintered HAp compacts and HAp coated Ti samples

Samples	Polarization condition	Sintering atmosphere	Maximum current density (nA/cm^2)	Stored charge (Qp) $(\mu C/cm^2)$
Bulk sintered HAp compact	$T_P = 400^{\circ}C$, $E_P = 2.0$ kV/cm, $t_P = 1$ h	Air	4.37	4.28
Hap coating	$T_P = 400^{\circ}C$, $E_P = 2.0$ kV/cm, $t_P = 1$ h	Air	1.19	1.69

Contact angle and surface energy data		
Samples	Contact angle (degree)	Surface energy (mJ/m^2)
Unpoled bulk sintered HAp compact	$65.43 \pm 0.73^{\circ}$	35.20 ± 0.89
'N' poled bulk sintered HAp compact	$48.04 \pm 0.0.39^{\circ}$	48.67 ± 0.87
'P' poled bulk sintered HAp compact	$47.7 \pm 0.36^{\circ}$	49.47 ± 3.67
Unpoled Hap coating	$6.77 \pm 0.51^{\circ}$	--
'N' poled HAp coating	$6.13 \pm 0.46^{\circ}$	--
'P' poled HAp coating	$5.97 \pm 0.67^{\circ}$	--

Figure 1. FESEM micrographs showing microstructure of (a) bulk sintered HAp compact and (b) heat treated HAp coating surfaces.

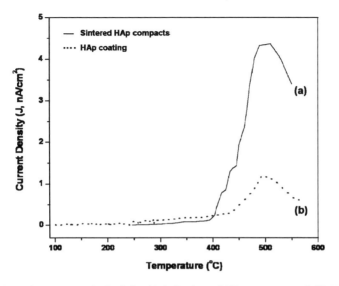

Figure 2. TSDC thermograms obtained for (a) bulk sintered HAp compact and (b) HAp coating samples polarized at 2.0 kV/cm for 1 h at 400°C

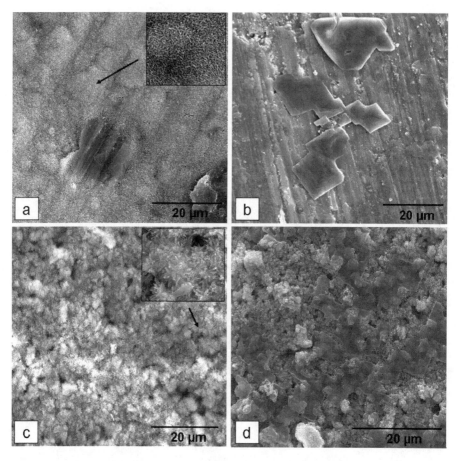

Figure 3. SEM micrographs of the samples surfaces after immersion in SBF for 5 days: (a) 'N' poled bulk sintered HAp compact, (b) 'P' poled bulk sintered HAp compact, (c) 'N' poled HAp coating and (d) 'P' poled HAp coating surfaces. High magnification images are shown in inset in Fig. 2a and 2b illustrating the morphology of the precipitated apatite layers. [Polarization condition: Samples were polarized at 400°C under an electric field of 2.0 kV/cm for 1 h].

FREEZE CASTING OF POROUS BIOMATERIAL SCAFFOLDS FOR BONE TISSUE ENGINEERING

Kajal K. Mallick
School of Engineering
University of Warwick
Coventry CV4 7AL
United Kingdom

ABSTRACT
This paper firstly reviews the current trends in fabrication techniques to produce porous inorganic biomaterial constructs for hard tissue augmentation. Of these, simple and cost effective freeze casting method has been employed over the last few years to process mainly bioceramics but has often met with difficulty in controlling variously oriented pore structures. The present work also compares two novel freeze casting methods using camphene and a mixture of water and glycerol in fabricating excellent networked 3-D porous structures of Bioglass and hydroxyapatite. It was possible to control the development of dendritic pore microstructure by the control and optimization of the process parameters such as mould temperature (0 to -196°C), temperature of freezing vehicle, glycerol concentration (5-40wt%) and temperature gradient driving the solidification and/or sublimation. DTA-TGA, XRD, density, and SEM analysis of bioscaffolds, air sintered up to 1100°C, showed monotonic correlations between porosity, pore size, orientation and solid loading (10-70%). A maximum porosity of 90% was achieved for structures with micro and macropores of 2-120μm, a clinically viable range that is amenable for bone in-growth and revascularization.

INTRODUCTION

In recent years, tissue engineering has emerged to overcome immunological, morbidity and disease transmission problems related to tissue repair such as grafting and other forms of tissue repair[1]. Over the last decade, significant advances have been made in the area of tissue engineering involving three dimensional (3-D) and interconnected porous scaffold constructs that are essential in the creation of new tissues from cultured cells[2-7]. This approach is now considered a commercially viable alternative route to hard tissue augmentation such as bone and teeth. Selection of appropriate synthetic or biologic (natural) biomaterials and especially fabrication methodologies are therefore critically important in designing temporary but synthetic extracellular scaffold matrices which can support the complex nature of tissue formation. The choice of biomaterials would depend on their specific resorbability so as to match the rate of new tissue growth over a certain period and may range from ceramics, glass and glass-ceramics to polymers as well as their composites. The characteristics of the pore architecture and both the mesoporosity (2-50 nm) and macroporosity (> 50μm) are vital as these provide a suitable platform for adhesion, differentiation and proliferation of cells that can effectively couple to vascularisation and delivery of metabolites. The motivation of the present work is therefore to present a brief overview of some of the current approaches available in producing porous bioscaffolds that are suitable for their use in bone tissue engineering[8]. Furthermore, results are reported for two freeze casting techniques developed in the present investigation namely, camphene freeze casting (CFC) and water and glycerol freeze casting (WGFC) in terms of their ability to produce networked 3-D porous structures using the well known Bioglass 45S5 and hydroxyapatite (HAP). A comparative analysis of the nature and influence of the freezing vehicle, pore structure, porosity and solid loading for the fabricated interconnected bioscaffold structures is also presented.

FABRICATION METHODS FOR POROUS SCAFFOLDS

For bone tissue engineering applications, there are principally two main methods i.e. rapid prototyping (RP) and conventional processing (CP), currently available for the fabrication of 3-D porous scaffold constructs using a variety of biomaterials such as alloys, ceramics and polymers. Some RP techniques can mimic the exact dimensionality of the desired tissue architecture. However, RP can often be time consuming, by and large expensive and, in some cases, are severely limited by the required spherical morphology of the feedstock together with lack of availability of some UV curable biodegradable polymer formulations, PLA and PGA, for example. RP methods are not reviewed[9] here but several conventional and economically feasible routes that can provide better control over the fabrication of the porous microstructures. Some of the most promising candidate methods are briefly reviewed here. These are: (1) Conventional sintering and Firing of ceramic powder compacts (2) Foaming (3) Gel-casting (3) Porogen or solvent leaching (4) Replication and (5) Freeze Casting

Conventional Sintering and Firing of Ceramic Powder with fugitive Phases

Synthetic porous structures are produced by these methods are popular. They are linked in that the conventional sintering involves sintering of the mixture of biomaterial and beads of polymer. Subsequent melting and vaporization of the polymer produces pores[10]. The latter is similar except a volatile or combustible pore forming fugitive phase is used to create space for the pores. In both cases, formation of regular interconnectivity often present difficulty due to irregular morphology of the feed or not being able to control the combustion kinetics.

Foaming

This generic technique of producing pores relies on the use of a foaming agent homogeneously dispersed in aqueous slurry of biomaterials or simply blowing gases through slurry. The method is unreliable, problematic to control the foaming process variables thus difficult to produce optimized pore structures.

Gel-Casting

This method, similar to slip casting, can produce complex-shaped profiles that are strong enough to be machined if necessary. A high content of ceramic or glass powders is suspended in an organic gel which polymerizes to form a cross-linked rigid green body. Normally, 10-20% of the slurry is used and the solvent is removed using drying as it sublimes. Although the process can be cost effective the negative issues are additive toxicity, control of the drying process and the difficult polymerization reactions.

Replication of Polymeric Sponge

This method, similar to slip casting, can produce complex-shaped profiles that are strong enough to be machined if necessary. A high content of ceramic or glass powders is suspended in an organic gel which polymerizes to form a cross-linked rigid green body. Normally, 10-20% of the slurry is used and the solvent is removed using drying as it sublimes. Although the process can be cost effective the negative issues are additive toxicity, control of the drying process and the difficult polymerization reactions.

Freeze Casting

Freeze casting technique is a process where a ceramic or glass slurry (usually aqueous) containing a certain vehicle is frozen in a non-porous mould. The frozen vehicle solvent acts as a binder and holds the slurry together as a green body with limited strength. The porogen acting as a vehicle is then removed by sublimation by freeze drying leaving an interconnected porous network. The green body is de-molded and sintered at appropriate temperatures to produce a porous body sample with improved mechanical strength.

EXPERIMENTAL

Materials and Scaffold Fabrication

The Bioglass 45S5 of the composition $24.5Na_2O.24.5CaO.45SiO_2.6P_2O_5$ in wt%, originally proposed by Hench, was melted in a 100g platinum crucible at 1450°C for 3 hours. The molten glass was cast, pulverized and micronized to a particle size < 4 um. The HAP, $Ca_{10}(PO_4)_6(OH)_2$, powder of 99.9% purity (Merck, Darmstadt, Germany) with a median particle size (d_{50}) of 3μm and a specific surface area of 72 m^2/g was used. Camphene, $C_{10}H_6$ (Sigma-Aldrich, UK), carboxylic acid (alkali-free and molecular mass of 320 g/mol Dolapix CE64, Zschimmer and Schwarz, Germany) and glycerol (Sigma-Aldrich, UK) were used as freezing vehicle, dispersant and cryoprotectant, respectively.

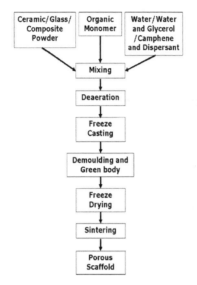

Figure 1. Freeze casting of porous bioscaffolds

A 40g batch of warm ceramic and Bioglass slurries, with solid loadings of 5-60 wt%, were first ball milled using a Dolapix concentration of 6 wt%, evacuated for 20 minutes to remove any trapped air bubbles. This was followed by casting in a stainless steel split mould which was cooled at temperatures from 0 to -196°C according to the particular freeze casting method employed. A flow chart of two freeze casting methods namely, camphene freeze casting (CFC) and glycerol and water freeze casting (GWFC), used to fabricate scaffold structures is shown in Figure 1. After careful removal from the mould followed by relevant freeze drying processing the green body was fired up to a maximum temperature of 1100°C.

Scaffold Characterization

Isothermal behavior of as-prepared slurries was determined by simultaneous differential thermal analysis and thermogravimetry (DTA-TGA) (STA1500 TA Instruments, West Sussex, UK). 20 mg of slurry was heated to 1000°C with a ramp rate of 10°C/min in flowing air.

Crystallinity and phase development of gold sputtered porous scaffolds were determined using powder X-ray diffraction (XRD) in the region of 2θ = 10-80° with a step size of 0.02°and step duration of 0.5s on a Ni-filtered Philips diffractometer (Model PW1710) using CuK_α radiation (λ = 0.15406 nm) at 40 kV and 40 mA. Using an automated powder diffraction software package the evolved phases were matched to both standard ICDD and calculated ICSD diffraction files.

Scanning electron microscopy (SEM) via a Philips Cambridge Stereoscan and JEOL Model 840 were used to study pore morphology as well as to observe related microstructural features of the porous architecture.

Open porosity and bulk density of the green body and the as-fired scaffolds were determined using the Archimedes method, pyconometry (Model AccuPyc II 1340, Micromeretics, UK), and mercury porosimetry (Autopore IV 9500, Micromeretics, UK). The porosity p of the scaffold was calculated by

$$p = 1 - \frac{\rho_{Scaffold}}{\rho_{Solid}}$$

where ρ_{solid} = 3156 kgm^{-3} for HAP and 2700 kgm^{-3} for Bioglass.

RESULTS AND DISCUSSION

Phase Transformation

DTA thermogram for the Bioglass exhibited the onset of the glass softening (T_g) at approximately 530°C with a peak endotherm due to the progressive glass transition at 560°C. Formation of new crystal phase is indicated by the onset of an exothermic crystallization (T_c) due to slow conversion to a glass-ceramic phase that begins at about 620°C with maxima at 670°C and finally the crystallization process is completed at 720-730°C. Melting (T_m) endotherm occurs at around 1050°C. These thermal events correspond well with those reported in the literature[11-13].

As expected, the XRD trace of the as-micronised Bioglass powders showed a broad 'halo' characteristic of an amorphous material. At 730°C, crystalline peaks were observed with increased peak intensity within what is still a broad background indicative of short range atomic order. This is evidence of the onset of just the surface rather than bulk crystallization of the Bioglass. The indexed peaks matched the standard JCPDS reference pattern (22-1455).for $Na_2Ca_2Si_3O_9$ (NCS), a phase reported also in other studies[14-18]. A similar study of a Na_2O and P_2O_5 free glass composition (70S30C) reported the formation of natural wallastonite (β-$CaSiO_3$). Presence of these phases is necessary since both the degree of crystallization and the kinetics of hydroxyl calcium apatite (HCA) serves as an indicator of bioactivity[14-16,19,21-22].

Pore Morphology

Camphene Freeze Casting (CFC): Development of pore architecture in Bioglass and HAP scaffolds, cast and freeze dried at -10°C, is shown in typical SEM micrographs in Figures 2 and 3, respectively. The observed porous microstructures are quite uniform throughout and the 3-D pore morphology is interconnected, degree of which varied with solid loading. The pores and interconnected nature of the fractured topography shown here are maintained for all samples although some pores had values in excess of 150μm. Replication of the unidirectional camphene sublimation resulted in the development of coralline-like pores and the struts. At 730°C, for Bioglass (Figure 2) in particular, this process is facilitated by a short temperature excursion at T_g thus controlling the limited viscous flow and eventual conversion to a surface crystallized glass-ceramic.

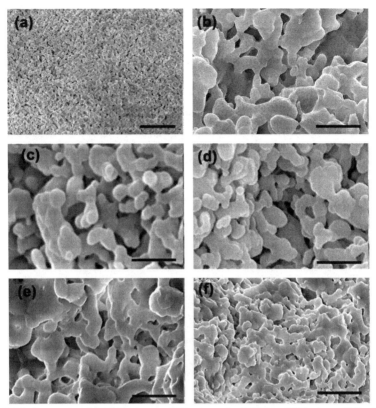

Figure 2. SEM micrographs of CFC Bioglass scaffolds sintered at 730°C at -10°C using solid loading of (a) General microstructure, (b) 10 wt%, (c) 20 wt%, (d) 40 wt%, (d) 60 wt% and (e) 70 wt% [(a) bar = 1mm ; (b), (c), (d), (e) bar = 200μm].

In general, the nature of the pore architecture for both Bioglass and HAP was found to be essentially similar even with solid loading as high as 70%. Some of the macropores observed in the range of around or above 150μm which exceeded the measured mean value of 120μm. These macropore values are the highest reported in the literature for camphene based method for Bioglass related compositions[23]. The microstructural development in this study is suggested to be the consequence of better control of the processing variables associated with the sublimation process. For HAP scaffolds, as shown in Figure 3 (e), the pores appeared to coalesce more with higher loading resulting in the consolidation of smaller pores at elevated temperature. This is to be expected as there is more aggregate contact at higher loading than those with low volume fraction. Other related work[17] has been reported on similar replicated macroporous bioglass foams.

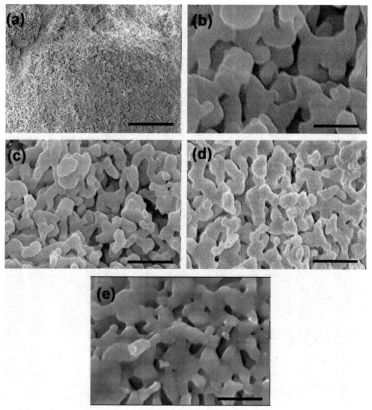

Figure 3. SEM micrographs of CFC HAP scaffolds sintered at 1100°C at -10°C using solid loading of (a) General microstructure, (b) 10 wt%, (c) 20 wt%, (d) 40 wt% and (e) 60 wt% [(a) bar = 1mm ; (b), (c), (d), (e) bar = 200μm].

Water and Glycerol Freeze Casting (WGFC): When a mixture of water and glycerol was used as a freezing medium and sublimed at -196°C the scaffold microstructures resembled those fabricated by CFC but with reduced level of porosity of maximum 80% and pore size of 60□m. Both the porosity and pore size are expected to increase with lowering the cooling and sublimation temperature. This was observed for the case of CFC but not carried out for this method. The porous morphology exhibited uniformly homogeneous and a dendritic geometry with thicker pore walls regardless of the solvent concentration. The interconnecting struts or pore walls were largely defect free. As the freeze casting process proceeds water molecules in the solvent act as a modifier thus eliminating associated defects. There was no evidence of the presence of large voids which, according to some authors[24,25], can result due to particle rejection. The general pore structure is reproducible and possible to control by monitoring the freeze drying temperature during casting and solvent sublimation.

Effect of Solid Loading on Porosity

For CFC derived scaffolds, the relationship of porosity with solid loading (10-60%) is monotonic, as shown in Figure 4. The porosity decreased with the increase in solid loading. A maximum porosity of 90% and 80% were achieved for Bioglass and HAP, respectively. In general, the uniform thickness of struts or interdendritic bridges, an indicator of pore interconnectivity, was on average 2-3µm for the lowest loading (10%) and this value increased to a maximum 15µm with increased level of loading. The variation of the measured mean pore diameter with loading is shown in Figure 5. The upper limit of the mean pore size for 10% loading Bioglass and HAP was around 120µm. A large difference in porosity at an intermediate range between 25 and 45% loading is noticeable between these two materials, which also exhibited a reduced level of microporosity of around 20µm at 60% and a further reduction to 10µm at 70% solid loading.

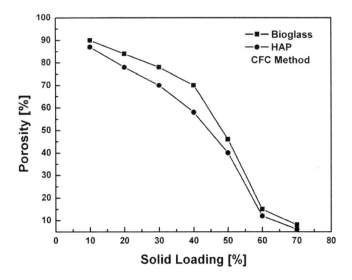

Figure 4. Measured porosity versus solid loading for CFC scaffolds at -10°C sintered at 730°C (■ Bioglass) and 1100°C (● HAP)

It was shown in previous studies[23] by the author that the mean pore diameter for Bioglass increases with the increase in sintering temperature due to ease with which glass flows thus producing more interconnected 3-D porous structures. However, in the present study, the temperature was maintained at a temperature, determined by DTA that corresponds to the end of the crystallization process, so as to specifically avoid the collapse of the interconnected scaffold network. The strategy here was to optimize the processing temperature to achieve an optimum but high level of pore size and interconnectivity. A similar trend was observed for HAP where, even with a high temperature of sintering at 1100°C, a lower level of maximum 85% porosity was achieved. However, mean pore size is comparable with Bioglass for similar loadings. The level of interconnectivity here is entirely dictated by the calescence of the ceramic particulates as sintering progresses.

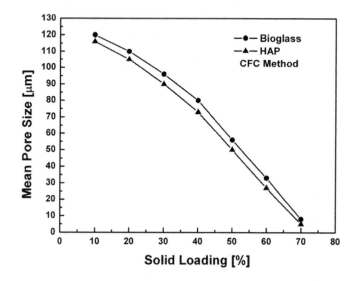

Figure 5. Plot of mean pore diameter of CFC scaffolds at -10°C as a function of solid loading sintered at 730°C (● Bioglass) and 1100°C (▲ HAP).

The relationship porosity and pore diameter for WGFC is nearly linear. However, as the scaffolds were cast and sublimed at liquid nitrogen temperature, the level of both values, as expected, are lower than those observed for CFC technique at -10°C. Still porosity values of 60-80% with 60-90μm pores were produced even with a medium level of loading up to 40 wt%. It was difficult to control the microstructure when the loading exceeded 50%. Further optimization work is ongoing to clarify the variables for process optimization. It is suggested that the key to achieve a homogeneous coralline geometry coupled to high level of porosity would be to find an optimized level of all three parameters namely, temperature, cryoprotectant concentration and the loading.

Influence of Freezing Temperature
    The effect of freezing temperature on the development of pore dimension in scaffolds fabricated using CFC for 20% solid loading is shown in Figure 7. The pore size distribution is unimodal for all four temperature regimes with peak mean pore diameter values around 80, 95, 120 and 130μm for liquid nitrogen (-196°C), dry ice (-78°C), -10°C and ice (0°C) as coolants, respectively. These results are in good agreement with microstructural observations of the samples.

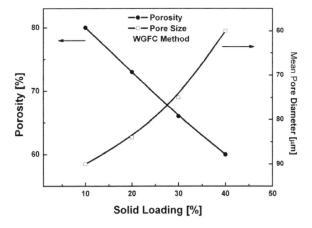

Figure 6. Plot of mean porosity and pore diameter for WGFC Bioglass scaffolds at -196°C as a function of solid loading heat treated at 730°C.

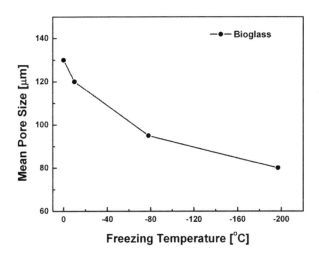

Figure 7. Plot of mean pore diameter versus freezing temperature for Bioglass scaffolds.

Influence of Sintering Temperature

The effect of sintering temperature and the variation in pore diameters with temperature for Bioglass and HAP are shown in Figure 8. The results shown are for a medium solid loading of 30%. At low sintering temperature of 550°C in Figure 8(a), corresponding to the onset of softening (T_g) the Bioglass, particulate clusters have not began to flow well enough to produce good links with the surrounding particles resulting in a porous structure that is not fully defined. In contrast, at the end crystallisation point of 730°C the particles flowed minimally, shown in Figure 8(b) so as to create, define and maintain a coralline interconnected porous texture. The thickness of pore walls varied from 8-10μm producing networked pores.

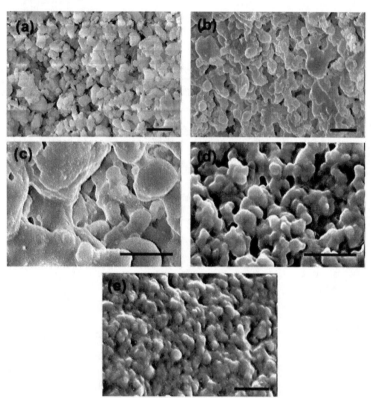

Figure 8. SEM micrograph showing the influence of sintering temperature for the CFC Bioglass scaffolds for a solid loading of 30 wt% at (a) 550°C (b) 730°C (c) 950°C and (d) 1050°C [(a), (b) bar = 100μm; (c), (d), (e) bar = 200μm].

As a comparison, at higher temperature of 950°C the pore channels, produced by camphene sublimation, are now more delineated but the pore channels opened up more producing a well lineated texture. At higher temperature of 1050°C the pores collapse in Figures 8(d) due to melting of the glass.

As the CFC technique induces only a thin surface crystallized layer (~3μm) of $Na_2Ca_2Si_3O_9$, the level of bioactivity is expected to be sufficient for resorption promoting biomineralisation and subsequent bone in-growth[14,22]. Crystallization behavior of this type has also been reported elsewhere[26]. An optimal sintering temperature of around 730°C for producing porous glass-ceramic scaffolds is achievable with even higher level of porosity than those reported recently[23] for this composition sintered at 850-950°C. The interconnected nature of the glass-ceramic is consistent in eliciting cell proliferation in implants resulting in better bone regeneration tailored to the rate of biodegradation[26-33].

For HAP, the effect of sintering temperature is reflected mainly in the slight reduction of porosity which is attributed to the grain coalescence or mass transport due to thermally activated atomic diffusion.

Influence of Freezing Vehicle

The relationship of porosity and pore size for HAP scaffolds with increasing glycerol concentration is monotonic (Figure 9). The results show that content of the cryoprotectant has marked effect on the development of pore structures and pore diameters in particular. Mean pore diameters varied from 10-57μm with a maximum of 55% porosity and a wall thickness of around 6-8μm. The smaller pores resulted from the effect of glycerol which reduces the size of the ice crystallites[25,34-39]. An average pore diameter of 55μm at 30% glycerol showing interconnected microporous surface pits (3-5μm) is considered beneficial for an accelerated cellular response[37]. A recent study reported[34] constructs of only 1-2μm pores with porosity of < 50% using this method. The freezing behavior is expected to be similar to that reported for constructs of alumina[25,38]. Adjustment of glycerol/water ratio is therefore clearly seen to be an optimizing factor in controlling macro/microporosity.

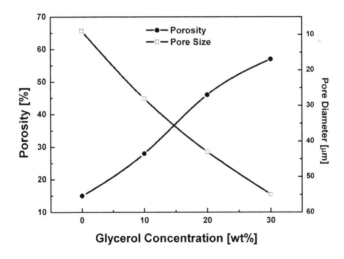

Glycerol Concentration [wt%]

Figure 9. Influence of solvent concentration on porosity and pore size of WGFC HAP scaffolds at 40 wt% solid loading (● Porosity and □ Pore size).

The bioresorbable structures with the presence of both macro and microporosity are ideal to accelerate the rate of HCA formation. The pore size range achieved will also satisfy the complex physiological factors required clinically. The architectural patterns, in the form of coralline/dendritic originated from the orientation along the solvent growth direction and hence porosity. This is will aid in the directional cell growth and rapid vascularisation of the implants. Also, the processes of both the osteoconduction and osteoinduction characteristics are expected to complement the observed and useful pore texture. SBF experiments are ongoing to assess the validity of these assertions and to assess the biological response of these resorbable materials.

CONCLUSION

CFC and WGFC methods with various slurry concentrations of 10-70 wt% demonstrated fabrication of porous three dimensional scaffolds of bioresorbabale Bioglass and HAP with maximum porosity of 90% and up to 150μm pore diameter. At low temperatures of 730°C, coralline-like uniform dendrites with micropores of 3-5μm on the interconnected bridges expected to promote in-vivo bioactivity. These are suitable for bone tissue engineering applications as the pore sizes exceeded 100μm considered as optimum range for bone in-growth and revascularization. It was found that lower the temperature of freezing smaller is the pore diameter. It is concluded that both methods are suitable with varying degree of control to fabricate structures which may be adapted for bone tissue engineering. Coalescence of particles at high temperatures resulted in reduction of pore diameter and porosity for both bioactive biomaterials.

REFERENCES

[1] R. Langer and J.P. Vacanti, Tissue engineering, *Science*, **260** 920–926 (1993).

[2] J.J Klawitter and A.M.Weinstein, The Status of Porous Materials to obtain Direct Skeletal Attachment by Tissue Ingrowth, *Acta Orthop. Belg.*, **40** 755-765 (1974).

[3] E.W. White, J.N. Weber, D.M. Roy, E.L. Owen, R.T. Chiroff and R.A. White, Replamineform Porous Biomaterials for Hard Tissue Implant Applications, *J. Biomed. Mater. Res.*, **9** 23-27 (1975).

[4] M. Spector, M.J. Michno, W.H. Smarook and G.T. Kwiatkowski, A High Modulus Polymer for Porous Orthopedic Implants: Biomechanical Compatibility of Porous Implants. *J. Biomed. Mater. Res.*, **12** 665–677 (1978).

[5] J.J. Klawitter, J.G. Bagwell, A.M. Weinstein and B.W. Sauer, An Evaluation of Bone Growth into Porous High density Polyethylene, *J. Biomed. Mater. Res.*, **10** 311-323 (1976).

[6] H.J. Cestero Jr, K.E. Salyer and I.R. Toranto, Bone Growth into Porous Carbon, Polyethylene, and Polypropylene Prostheses, *J. Biomed. Mater. Res.*, **9** 1-7 (1975).

[7] B.W. Sauer, A.M. Weinstein, J.J. Klawitter, S.F. Hulbert, R.B. Leonard and J.G. Bagwell, The Role of Porous Polymeric Materials in Prosthesis Attachment, *J. Biomed. Mater. Res.*, **8** 145-153 (1974).

[8] C.W. Patrick, Jr, A.G. Mikos and McIntire, Prospectus of Tissue Engineering In: Frontiers in Tissue Engineering, Editors: C.W. Patrick, Jr, A.G. Mikos and L.V. McIntire, Elsevier Science, New York, USA, 3–14 (1998).

[9] Scott J. Hollister, Porous Scaffold Design for Tissue Engineering, *Nature Materials*, 4, 518-524 (2005).

[10] A. Uchida, S.M. Nade, E.R. McCartney and W. Chiang, The Use of Ceramics for Bone Replacement. A comparative study of three Different Porous Ceramics, *J. Bone Joint Surg. Br.*, **66** 269-275 (1984).

[11] H.A. Batal, M.A. Azooz, E.M.A. Khalil, A. Soltan Monem and Y.M. Hamdy, Characterization of Some Bioglass–ceramics, *Mater. Chem. Phys.*, **80 [3]** 599–609 (2003).

[12] A. El Ghannam, E. Hamazawy and A. Yehia, *J. Biomed. Mater. Res.*, Effect of Thermal Treatment on Bioactive Glass Microstructure, Corrosion Behavior,Potential, and Protein Adsorption, **55 [3]** 387–398 (2001).

[13]X. Chatzistavrou, T. Zorba, E. Kontonasaki, K. Chrissafis, P. Koidis and K.M. Paraskevopoulos, Following Bioactive Glass Behavior Beyond Melting Temperature by Thermal and Optical Methods, *Phys. Stat. Sol.(a)*, **201** [5] 944–951 (2004).

[14]D.C. Clupper, J.J. Mecholsky Jr, G.P. LaTorre, D.C. Greenspan, Sintering Temperature Effects on the In Vitro Bioactive Response of Tape Cast and Sintered Bioactive Glass–ceramic in Tris Buffer, *J. Biomed. Mater. Res.*, **57** [4] 532–40 (2001).

[15]D.C Clupper, J.J Mecholsky Jr., G.P LaTorre, D.C Greenspan, Bioactivity of Tape Cast and Sintered Bioactive Glass–ceramic in Simulated Body Fluid, *Biomaterials*, **23** [12] 2599–2606 (2002).

[16]O.P. Filho, G.P. LaTorre, L.L. Hench, Effect of Crystallization on Apatite-layer Formation of Bioactive Glass 45S5, *J. Biomed. Mater. Res.*, **30** 509 (1996).

[17]Q.Z. Chen, I.R. Thomson and A.R. Boccaccini, 45S5 Bioglass® – Derived Glass-ceramic Scaffold for Bone Tissue Engineering, *Biomaterials*, **27** [11] 2414-25 (2006).

[18]I.K. Jun, Y.H. Koh and H.E. Kim Fabrication of Highly Porous Bioactive Glass-ceramic Scaffold with High Surface Area and Strength, *J. Amer. Ceram. Soc.*, **89** [1] 391-4 (2006).

[19]P. Li, Q. Yang, F. Zhang and T. Kokubo, The Effect of Residual Glassy Phase in a Bioactive Glass-ceramic on the Formation of its Surface Apatite Layer In Vitro, *J. Mater. Sci : Materials in Medicine*, **3** [6] 452–56 (1992).

[20]O. Peitl, E.D. Zanotto and L.L. Hench, Highly Bioactive P_2O_5–Na_2O–CaO–SiO_2 Glass Ceramics, *J. Non-Cryst. Solids*, **292** [1-3] 115–26 (2001).

[21]D. C. Clupper and L. L. Hench, Crystallization Kinetics of Tape Cast Bioactive Glass 45S5, *J. Non-Cryst. Solid*, **318** [1-2] 43-8 (2003).

[22]O. Peitl, G.P. LaTorre and L.L. hench, Effect of Crystallisation on Apatite-Layer Formation of Bioactive Glass 45S5, *J. Biomed. Mater. Res.*, **30** [4] 509-14 (1996).

[23]Kajal K. Mallick, Freeze Casting of Freeze Casting of Porous Bioactive Glass and Bioceramics, *J. Amer. Ceram. Soc.*, **92** [S1] S85-S94 (2009).

[24]K.A.Keler, G.M. Mehrotra and R.J. Kernas, Freeze Forming of Alumina Monoliths: pp. 557-67, in Processing and Fabrication of Advanced Materials. V. Edited by T.S. Srivatsan and J.J. Moore, Minerals, Metals and Materials Society/AIME, Warrendale, PA, (1996).

[25]S.W. Sofie and F. Dogan, Freeze Casting of Aqueous Alumina Slurries with Glycerol, *J. Amer. Ceram. Soc.*, **84** [7] 1459-64 (2001).

[26]L.L. Hench, Sol-gel materials for bioceramic applications, *Current Opinion Solid State Mater. Sci.*, **2** 604-10 (1997).

[27]A.E. Clark and L.L. Hench, Calcium Phosphate Formation on Sol-gel Derived Bioactive Glasses *J. Biomed. Res.*, **28** 693-8 (1994).

[28]L.L. Hench and J. Wilson, Surface Active Biomaterials, *Science*, **226** 630-6 (1984).

[29]L.L. Hench, R.J Splinter, W.C Allen, T.K Greenlee, *J. Biomed. Mater. Res. Symp.*, **2** [1] 117-41 (1971).

[30]W. Cao and L.L. Hench, Bioactive Materials, *Ceram. Int.*, **22** 493–507 (1996).

[31]T. Kokubo, Novel Bioactive Materials, *Anales de Qumica*, **93** S49-55 (1997).

[32]V. Banchet, E. Jallot, J. Michel, L. Wortham, D. Laurent-Maquin and G. Balossier, *Surf. Interface Anal.*, X-ray Microanalysis in STEM of Short-term Physicochemical Reactions at Bioactive Glass Particle/Biological Fluid Interface. Determination of O/Si Atomic Ratios, **36** [7] 658–65 (2004).

[33]T. Kokubo, H. Kushitani and S. Sakka, *J. Biomed. Mater. Res.*, Solutions able to Reproduce in vivo Surface-structure Changes in Bioactive Glass-ceramic A-W3, **24** [6] 721-34 (1990).

[34]Q.Fu, M.N. Rahaman, F Dogan and B.S. Bal, Freeze Casting of Hydroxyapatite Scaffolds for Bone Tissue Engineering Applications, *Biomedical Materials*, IOP Publishing UK, **3** 1-7 (2008).

[35]B-H. Yoon, Y-H. Koh, C-S. Park, and H-E. Kim, "Generation of Large Pore Channels for Bone Tissue Engineering Using Camphene-Based Freeze Casting, *J. Am. Ceram. Soc.*, **90 [6]** 1774-752 (2007).
[36]E.R. Rubenstein and M.E. Glicksman, Dendritic Growth Kinetics and Structure II. Camphene, *J. Crys. Growth*, **112** 97-110 (1991).
[37]K. Anselme, Osteoblast Adhesion in Biomaterials, *Biomaterials*, **21 [7]** 667-81 (2000).
[38]K. Lu, C.S. Kessler and Richey M. Davis, Optimization of a Nanoparticle Suspension for Freeze Casting, *J. Am. Ceram. Soc.*, **89 [8]** 2459-65 (2006).
[39]Q.Fu, M.N. Rahaman, F Dogan and B.S. Bal, Freeze Casting of Porous Hydroxyapatite Scaffolds: I. Processing and General Microstructure Solutions, *J. Biomed. Mater. Res. Part B: Appl. Biomater.* (2008), DOI: 10.1002/jbm.b30997

To be presented at the 8th Pacific Rim Conference on Ceramic and Glass Technology (PACRIM 8), Vancouver, British Columbia, CANADA, May 31- June 5, 2009
[*]Member, American Ceramic Society
[**]Author to whom correspondence should be addressed
email: k.k.mallick@warwick.ac.uk

HARD TEMPLATE SYNTHESIS OF MESOPOROUS HYDROXYAPATITE MATERIALS FOR CONTROLLED PROTEIN RELEASE

Janina Möller, Philippe Dibandjo, Roger Gadiou, Joseph Dentzer, Karine Anselme, and Cathie Vix-Guterl
IS2M Institut de Science des Matériaux de Mulhouse, LRC CNRS 7228 UHA
15, rue Jean Starcky, BP 2488, F-68057 Mulhouse, France

ABSTRACT
Calcium phosphates are of prime importance in the field of hard tissue repair and orthopedic implants due to their high biocompatibility and -resorption.
The aim of the present work is to develop an injectable bone filling system based on calcium phosphates possessing controlled nanoporosity and allowing controlled release of growth factors in situ. In comparison to macroporous materials, it can be expected that nanoporous calcium phosphates show better biomolecule retention combined with improved growth factor release control. Our objective is to synthesize a nanoporous material having pores size between 2-30 nm, big enough for protein immobilization.
To fulfill this goal, new synthesis strategies must be proposed. Conventional procedures used to prepare hydroxyapatite (HA) lead to porous calcium phosphates with micro- (< 2 nm in diameter) or macroporosity (> 50 nm). In this work, we propose new synthesis pathways, based on templating techniques with the use of organized porous silica (SBA-15 or mesocellular foam) or ordered porous carbon as template, filled with aqueous HA or calcium metaphosphate precursors. The template elimination by oxidative heat treatment (carbon template) or by etching with NaOH (silica template) resulted in a porous calcium phosphate material with a high surface area and an interconnected porosity. A multi-scale characterization was performed on the obtained ceramics.
The structural and textural characteristics of the HA obtained will be discussed in relation with the synthesis conditions and the nature of the host material used.

INTRODUCTION
The increasing life expectancy of the populations in developed countries triggers rising need for cures to aged related diseases. Within this field, bone regeneration is one of the most important challenges for future generations. In most cases, patients are subjected to risky orthopedic interventions replacing the ill bone with metal or ceramic implants. The injection of third generation bone fillers can be an alternative to surgery reducing the mortality of the patients. Third generation biomaterials for bone regeneration are intended to activate auto-regeneration of the bone by providing ions, proteins and even cells at the focused site inside the bone[1].
The aim of the present work is to develop an injectable bone filling system based on calcium phosphates possessing controlled nanoporosity and allowing controlled release of growth factors in situ. In comparison to macroporous materials, it can be expected that nanoporous calcium phosphates show better biomolecule retention combined with improved growth factor release control. Our objective is to synthesize a mesoporous material having pores size between 2-30 nm, big enough for protein immobilization.
To fulfill this goal, new synthesis strategies must be proposed. Conventional procedures used to prepare hydroxyapatite (HA) lead to porous calcium phosphates with micro- (< 2 nm in diameter) or macroporosity (> 50 nm)[2]. In this work, we propose new synthesis pathways, based on templating techniques with the use of organized porous silica (mesocellular foam) or ordered porous carbon as template, filled with aqueous hydroxylapatite precurseurs.

SYNTHESIS AND CHARACTERIZATIONS

Silica foams were obtained by condensing tetraorthosilicate in acidic solution in presence of anionic surfactant Pluronic 123 and the swelling agent mesitylene adapted from Schmidt-Winkel et al.[3]: 10 g of Pluronic 123 are dissolved in 375 mL of 1.6 molar hydrochloric acid on an oil bath heated to 40°C. 5g of mesitylene are added and stirred for 30 minutes. 22 g of tetraorthosilicate are added and stirred at 40°C for 24h. The mixture is aged without stirring at 110°C for 24 h in a sealed container. The resulting solid is filtered, but not washed and dried at 80°C overnight. Calcination is carried out under air flow (20 L/h) at 500°C for 8 h. Heating ramp was 2°C/min. The silica is characterized by nitrogen physisorption experiments, small angle X-ray diffraction and transmission electron microscopy.

Carbon templates are obtained by chemical vapour deposition (CVD) of propylene at 750°C: 1.2 g of silica are introduced into a vertical oven and heated under low argon flow (10 L/h) to 750°C with a ramp of 10 °C/min. Once the temperature reached, the argon flow is increased to 20 L/h and a propylene flow is introduced (0.5 L/h). After 16 h of deposition, the composite is allowed to cool down under argon flow (10 L/h). The amount of carbon deposited can be determined by the mass difference after selective oxidation in air at 750 °C. The carbon template is retrieved after HF etching of the silica: The powder is treated at room temperature with HF solution (32 % in water) for at least 4h. The product is filtered, washed several times with distilled water and dried at 80 °C overnight. Characterization is carried out by nitrogen physisorption and transmission electron microscopy.

Hydroxyapatite powders are made by replication of silica or carbon templates. The amount of hydroxyapatite gel is calculated in function of the pore volume of the template. 1 g of carbon or silica template is degassed overnight under vacuum at room temperature. The valve to the vacuum ramp is closed and a solution of 16.92 g calcium nitrate hexahydrate in 15 mL distilled water is added via a septum with a syringe under stirring. The mixture is allowed to come back to atmospheric pressure. 7.46 mL of triethyl phosphine prehydrolyzed overnight with 0,1 mL of distilled water are added. The solution and the solid are stirred for 30 minutes, then aged without agitation at 60°C for one hour.

The liquid is evaporated at atmospheric pressure at 80°C at least overnight. The template removal technique depends on the template nature: the carbon template is removed by oxidative heat treatment under air flow (20 L/h) for 8h with a ramp of 5°C/min at 750°C. The silica template is removed by refluxing with 1 molar sodium hydroxide solution in ethanol and water (1:1 in volume) for 2h. The product is characterized by nitrogen physisorption, wide angle X-ray diffraction and transmission electron microscopy.

In order to study the porosity of the materials, nitrogen physisorption experiments were performed at 77K on a Micromeritics ASAP 2020 apparatus. The samples were first degassed overnight at 150°C. Adsorption and desorption isotherms were then recorded for relative pressures P/P_0 ranging between 10^{-3} and 1. The specific surface area of the material was obtain by applying the BET equation on the adsorption isotherm in the relative pressure range [0.05 – 0.30]. The pore size distribution was obtained with the BJH method[4]; this model based on Kelvin equation allows the description of capillary condensation of liquid nitrogen in mesopores. It is suitable for pores which have a size higher than 3 nm. The method was applied on the desorption isotherm for relative pressures between 0.9 and 0.3.

Wide angle X-ray diffraction was carried out on Philips X'Pert with Copper K_α radiation ($\lambda = 1.54060$ Å). Small angle X-ray diffraction was performed on a Rigaku SAXS between 0.03 and 4.42°. Transmission electron microscopy was executed on a Philips CM 200 with an acceleration voltage of 200 kV.

RESULTS AND DISCUSSION

Mesoporous silica is obtained after calcination. The ordering is less than for hexagonal silicas such as SBA-15, but the pore size distribution is small (see figure 1). The specific surface area is 751 m²/g and the mean pore size computed from the BJH pore size distribution is 6 nm, the specific pore volume was calculated to 1.8 cm³/g. Transmission electron micrographs show a highly porous structure with interconnected spherical pores.

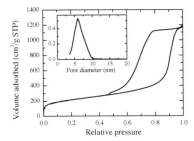 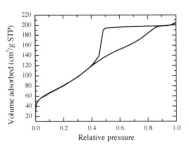

Figure 1. Nitrogen physisorption isotherms for silica (left) and carbon template (right). The insert shows the narrow pore size distribution in function of the pore volume for the silica template.

The carbon template obtained after HF etching is highly porous (pore volume 0.46 cm³/g) with specific surface area of 356 m²/g. In favor of high porosity, the silica template pores are not completely filled during the CVD by limiting the deposition time to 16 h. Longer deposition leads to better replication of the silica pore structure but results in carbon templates with less porosity available for the filling with the hydroxyapatite sol. Figure 3 compares transmission electron micrographs of the silica template, the carbon replica and the hydroxyapatite obtained from both templates.

White calcium phosphate powders were obtained from silica templates. Products from carbon templates were slightly gray showing some residual carbon. X-ray diffraction patterns confirm that the product is composed of pourly crystallized hydroxyapatite (JCPDS 09-0432, not shown.). As the silica templated product was not subjected to calcination, its cristallinity is lower than the one of the carbon templated one.

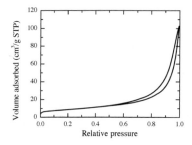 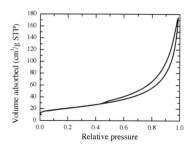

Figure 2. Nitrogen physisorption isotherms for silica (left) and carbon (right) templated hydroxyapatite powders

Specific surface areas range from 14.08 m²/g (carbon template) to 72,4m²/g (silica template). The shape of the nitrogen adsorption isotherm hysteresis (type H3) is typical for interparticular porosity.[5] These values are similar to those of hydroxyapatite obtained by classic sol-gel processes and suitable for protein adsorption. The higher surface area of the silica templated product can be explained by the absence of a calcination step. During this step, for the carbon templated product, grain growth is induced by the heat treatment resulting in less surface area. Consequently the grains of the silica templated ceramic are smaller (see transmission electron micrographs, figure 3). Both powders present intergranular porosity in which proteins can be adsorbed.

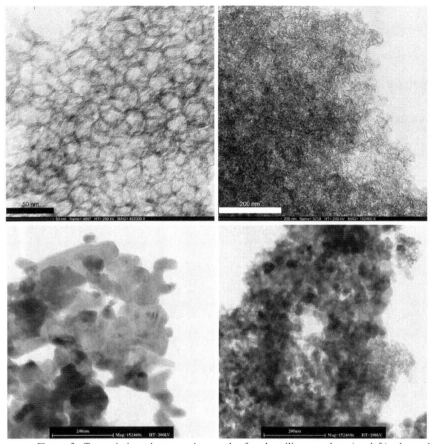

Figure 3. Transmission electron micrographs for the silica template (up left), the carbon template (up right) and the hydroxyapatite templated by silica (down left) and carbon (down right).

CONCLUSION

The described templating approach leads to hydroxyapatites with considerable surface areas and intergranular porosity suitable for protein adsorption. Both silica and carbon templates allow the synthesis of ceramics with interesting structural features. Protein adsorption experiments and tests for in vitro biocompatibility are currently in progress.

ACKNOWLEDGEMENT

The financial support by the ANR project PNANO NANOBONEFILLER (ANR-06-NANO-022) is gratefully acknowledged by the authors.

REFERENCES

[1]S.V. Dorozhkin, *J. Mater Sci* **44** 2343-87 (2009)
[2] S. Sanchez-Salcedo, A. Nieto, M.Vallet-Regi, *Chem. Eng. J.* **137** 62 (2005)
[3] P. Schmidt-Winkel, W.W. Lukens Jr, P. Yang, D.I. Margolese, J.S. Lettow, J.Y. Ying, G.D. Stucky, *Chem. Mater.* **12** 686-96 (2000)
[4] Rouquerol, F.; Rouquerol, J. & Sing, K. "Adsorption by powders and porous solids - principles, methodology and applications" chapter 7 Academic Press, London, 1999
[5]James B. Gondon, Surface Area and Porosity Determination by Physisorption, Elsevier Science, 1st edition (2006)

IMMOBILIZATION OF HEPARIN ON GELATIN MODIFIED THREE-DIMENSIONAL OSTEOCONDUCTIVE Ca-P/PHBV NANOCOMPOSITE SCAFFOLDS

Bin Duan, Min Wang[*]
Department of Mechanical Engineering, The University of Hong Kong
Pokfulam Road, Hong Kong
Email: memwang@hku.hk

ABSTRACT
        In bone tissue engineering, an effective and sustained delivery of growth factors such as bFGF, VEGF and BMP-2 at the target site is of significant importance. Heparin, a sulfated polysaccharide, is known to have the binding affinity with a number of growth factors and thus capable of blocking their degradation and prolonging their release time. In this study, the surface modification of three dimensional Ca-P/PHBV scaffolds produced via selective laser sintering was achieved through physical entrapment of gelatin by using a miscible mixture of a solvent and a non-solvent for PHBV, with gelatin being dissolved in the mixture. The amount of gelatin entrapped on the surface of scaffold struts was determined quantitatively by bicinchoninic acid (BCA) kit assay. After surface modification, a hydrophilicity/hydrophobicity balance was obtained for the surface of scaffold struts. Heparin was subsequently surface immobilized through covalent conjugation onto gelatin-modified scaffolds. The amount of heparin conjugated to the scaffold surface was determined using the toluidine blue method. Ca-P/PHBV scaffolds with entrapped gelatin and immobilized heparin should have enhanced binding of growth factors and hence promote cell proliferation and bone tissue regeneration.

INTRODUCTION
        World-wide, bone tissue regeneration using porous biodegradable scaffolds produced by different techniques has been extensively investigated. Among various scaffold fabrication techniques, solid free-form fabrication (SFF) or alternatively named "rapid prototyping (RP)", which represents a technology for a group of manufacturing techniques that can generate, in a layer-by-layer manner, a physical model directly from computer-aided design data, is considered to be a promising technology for scaffold fabrication because it can achieve extensive and detailed control over scaffold architecture[1,2]. Selective laser sintering (SLS) is one of the widely used RP techniques. On the other hand, an effective and sustained delivery of growth factors such as basic fibroblast growth factor (bFGF), vascular endothelial growth factors (VEGF) and bone morphogenetic proteins-2 (BMP-2) at the target site is of significant importance due to their ability to regulate various cell functions including adhesion, growth, secretion, apoptosis and differentiation[3]. Although growth factors play important roles in harnessing and controlling cellular functions in tissue regeneration, their short half-life, rapid degradation rate, relatively large size, slow tissue penetration and their potential toxicity at systemic levels leading to a long time for tissue to respond, limit their direct therapeutic applications[4]. Therefore, the carrier and delivery system for the sustained delivery of growth factors in a temporal and spatial way are very important for bone regeneration. Heparin, a sulfated polysaccharide, is known to have the binding affinity with a number of growth factors and thus capable of blocking the degradation of protein and prolonging the release time[5,6]. It was shown that in the presence of heparin, the degradation of BMP-2 was blocked, and the half-life of BMP-2 in the culture

medium was prolonged by nearly 20-folds[7]. Heparin conjugated scaffolds could also increase the ectopic bone formation, indicating that the sustained and controlled delivery of BMP-2 increased the bone regenerative efficacy of BMP-2[8,9]. In addition, the incorporation of heparin onto biomaterials could enhance biocompatibility, cell adhesion and growth[10,11].

In our previous investigations, three-dimensional calcium phosphate (Ca-P)/ poly(hydroxybutyrate-co-hydroxyvalerate) (PHBV) scaffolds were fabricated from Ca-P/PHBV nanocomposite microspheres using the SLS technique[12]. In the present study, the surface modification of Ca-P/PHBV nanocomposite scaffolds was conducted by the physical entrapment of gelatin on scaffold surface. In order to further improve the growth factor binding ability, heparin was immobilized on the surface of gelatin-modified scaffolds through covalent conjugation. The structure and properties of Ca-P/PHBV scaffolds thus modified were investigated.

MATERIALS AND METHODS

Fabrication of Ca-P/PHBV nanocomposite microspheres

Ca-P nanoparticles were made in-house by rapid mixing of $Ca(NO_3)_2 \cdot 4H_2O$ acetone solution with $(NH_4)_2HPO_4$ aqueous solution[13]. With sizes in the range of 10-30 nm, the Ca-P nanoparticles produced were amorphous and had a Ca:P molar ratio of about 1.5. Nanocomposite microspheres of Ca-P/PHBV (with the PHBV having a 3-hydroxyvalerate content of 12 mol%) were fabricated using a solid-in-oil-in-water (S/O/W) emulsion solvent evaporation method. The Ca-P/PHBV nanocomposite microspheres had an average diameter of 45.36 μm and a Ca-P content of 12.93 wt.%.

Three dimensional scaffold design and fabrication

A porous scaffold model with a three-dimensional periodic porous architecture, as shown in Fig.1, was designed using SolidWorks®. To facilitate scaffold handling during SLS and subsequent experiments, a solid base was incorporated in the scaffold design to provide a firm foundation for the scaffold. The design was then exported into an STL format and transferred to a modified Sinterstation® 2000 system (3D Systems, Valencia, CA) for making scaffolds using the SLS technique. According to our optimization results for scaffold fabrication via SLS, the part bed temperature for SLS in the current study was fixed to be 35 °C in order to prevent caking and the value of roll speed and scan speed were set to be their default values, 127 mm/s and 1257 mm/s, respectively. The laser power, scan spacing and layer thickness were set to be 15 W, 0.10 mm and 0.10 mm, respectively.

Surface entrapment of gelatin

Gelatin (Sigma, USA) was dissolved in a miscible mixture of 2,2,2-trifluoroethanol (TFE) and water (30:70 v/v) which are solvent and non-solvent for PHBV, respectively. The sintered Ca-P/PHBV scaffolds (two layers, ~70 mg) were immersed for 6 h in the gelatin solution whose concentration was 5 mg/ml. After this treatment, the scaffolds were rinsed in distilled water for three times to remove the non-entrapped gelatin and then dried at room temperature. The effects o TEF concentration and immersion time on the amount of gelatin entrapped were also investigated. The amount of gelatin entrapped on the surface of scaffold struts was determined quantitatively by bicinchoninic acid (BCA) kit assay[14].

Figure 1. Designed scaffold model using SolidWorks® (left) and a Ca-P/PHBV nanocomposite scaffold produced via SLS (right).

Surface immobilization of heparin

For heparin immobilization, gelatin entrapped Ca-P/PHBV scaffolds were firstly immersed in MES buffer solution (0.1 M, pH = 5.6) for 30 min at room temperature for pre-wetting. 2 mg of heparin (Mw = 17,000, activity 170USP units/mg; Sigma) were dissolved in a 1 ml MES buffer solution containing 2 mg N-hydroxysuccinimide (NHS; Sigma) and 1.2 mg 1-ethyl-3-(3-dimethylaminopropyl)-carbodiimide hydrochloride (EDC; Sigma). After 4 h activation at room temperature, the pre-wetted scaffolds were soaked into the activated heparin solution in MES buffer for another 4 h and then extensively washed with PBS and dried over night at room temperature. The amount of heparin conjugated to the surface of gelatin entrapped scaffold was determined using the toluidine blue method[15]. Heparin conjugated scaffolds were added to 1 ml 0.005% toluidine blue solution containing 0.2% NaCl and 0.01 N HCl. The mixture was thoroughly agitated with a vortex mixer and incubated at room temperature for 3 h. 1 ml of n-hexane was then added into the solution, vigorously mixed by vortex and centrifuged at 5000 rpm for 5 min. After removing n-hexane, absorbance at 631 nm was measured using a microplate reader. The calibration curve was constructed based on experiments using a series of heparin concentrations. The stability of immobilized heparin on scaffolds was studied by immersing heparin conjugated scaffolds in PBS solution at 37 °C for up to 14 days and the remaining heparin was qualitatively measured using the toluidine blue method.

Characterization

The morphology of scaffolds, before and after surface modifications, was examined using scanning electron microscopy (SEM). The porosity of the scaffolds was measured using a density kit and an electronic balance based on Archimedes principle. Due to difficulties caused by the high porosity of the porous scaffolds, the hydrophilicity of the scaffold surface before and after surface modifications was studied through contact angle measurements by using solvent-cast films instead of sintered scaffolds. Contact angles were measured at room temperature using the sessile drop technique, which was conducted on a contact angle measuring machine equipped with an image analysis tool.

Statistical analysis

All experiments were performed in triplicate. The data obtained are presented as means±standard deviation (SD). Statistical analyses were performed using ANOVA (analysis of variance) with a Scheffé test. Differences were considered statistically significant at *$p<0.05$ and **$p<0.01$.

RESULTS AND DISCUSSION

Ca-P/PHBV nanocomposite microspheres were produced in-house using the S/O/W emulsion solvent evaporation method. They were subsequently used for fabricating Ca-P/PHBV nanocomposite scaffolds via SLS. The successful modification of the Sinterstation® 2000 system in our previous investigation resulted in: (a) the consumption of biomaterial powders for scaffold construction was significantly reduced; and (b) good quality scaffolds were made. As shown in Fig.1, the designed scaffold model had eight layers. The distance between each layer and each pillar was 0.8 mm and the solid pillars were 1 mm in diameter. During the initial study of SLS for scaffold fabrication, it was found that the first several layers tended to contract significantly due to the initial input of thermal energy, which affected the overall quality of sintered scaffolds. To solve this problem, a solid base of the dimensions of $8.4 \times 8.4 \times 3$ cm^3 was thus incorporated. A successfully sintered Ca-P/PHBV scaffold is shown in Fig.1.

Fig.2(a) and 2(b) were top views of the designed scaffold model and actual, sintered Ca-P/PHBV nanocomposite scaffold. The SEM image of the sintered scaffold indicated that scaffold fabrication using SLS was successful as the scaffold morphology was very similar to the design and pores in the scaffold were clearly identified and comparable to the design. The high-magnification SEM images of the strut surface, as shown in Fig.2(c), showed that there was necking among adjacent microspheres, as well as intact microspheres without apparent fusion. The intact microspheres adhered to sintered scaffolds due to the heat generated during SLS and resulted in micropores between microspheres on the scaffolds. Therefore, through measurements, the porosity of sintered scaffolds was determined to be 61.75±1.24%, which is higher than the theoretic value (53.49%) of the designed scaffold model.

For the surface modification, physical entrapment of gelatin was firstly employed, which involved the use of a miscible mixture of solvent and non-solvent for PHBV. During the entrapment process, PHBV matrix of the Ca-P/PHBV struts swelled but did not dissolve in the gelatin solution. Gelatin molecules diffused into the swollen surface of PHBV and were entangled with PHBV molecules on the strut surface. After the scaffold was removed from the gelatin solution and immersed in water, which is a non-solvent for the polymer, the surface rapidly shrank and gelatin molecules on the polymer surface were entrapped and immobilized. This physical entrapment method was also used by others for the surface modification of their tissue engineering scaffolds[16].

The entrapped gelatin was measured using the BCA method, which is an effective way for determining the amount of the immobilized protein on the surface of a biomaterial. For two-layer Ca-P/PHBV scaffolds used in the current study, the effects of TFE concentration and immersion time on the amount of entrapped gelatin are shown as the two diagrams in Fig.3. When the TFE concentration was 0%, which means that the scaffolds were simply immersed in aqueous gelatin solution, only 823.65±78.40 μg gelatin/scaffold was absorbed onto the surface of the scaffold struts instead of physical entrapment. The amount of entrapped gelatin significantly increased with increasing TFE concentration, which was caused by the increase of swelling of PHBV. When the TFE concentration was increased to 50%, the scaffolds became too soft to be handled due to the high amount of solvent (TFE) and hence the entrapped gelatin could not be measured. As shown in Fig.3(b), with an increase in immersion time, i.e., the reaction time, the amount of entrapped gelatin increased significantly. However, with increasing immersion time, mechanical properties of the scaffolds decreased due to polymer matrix swelling. Therefore, the treatment condition of 30% TFE and 6 h immersion time was adopted for Ca-P/PHBV scaffolds that would be used for subsequent heparin immobilization.

(a)

(b)

(c)

(d)

Figure 2. Top view of the designed scaffold model (a) and SEM image of a sintered Ca-P/PHBV nanocomposite scaffold, and high magnification SEM images of Ca-P/PHBV scaffolds before surface modification (c) and after heparin immobilization (d).

After gelatin entrapment on the surface of Ca-P/PHBV scaffolds, heparin was covalently bonded to amine groups of the entrapped gelatin molecules, as in standard carbodiimide chemistry, using the aforementioned experimental procedure. The amount of immobilized heparin was determined to be 41.78±0.39 μg heparin/scaffold. The high magnification SEM image of a Ca-P/PHBV scaffold after heparin immobilization is shown as Fig.2(d). It was observed that some Ca-P/PHBV microspheres lost their original rough-wrinkled morphology due to the swelling of PHBV during the physical entrapment process.

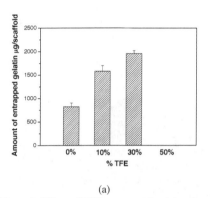 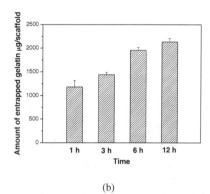

(a)   (b)

Figure 3. Effects of TFE concentration (a) and entrapment time (b) on the amount of entrapped gelatin. Any two data points are significant different from each other ($^{**}p<0.01$) in respective diagrams.

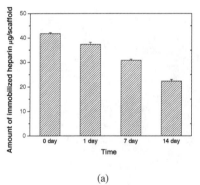 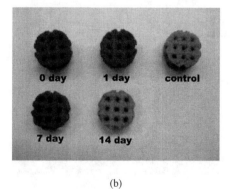

(a)   (b)

Figure 4. (a) The amount of immobilized heparin changed with immersion time. (b) Image of heparin immobilized scaffolds after toludine blue staining at different immersion time point.

The stability of immobilized heparin on Ca-P/PHBV scaffolds is very important because it can affect the binding stability and release profile of BMP-2 when BMP-2 is attached to the scaffolds. Fig.4(a) displays the changes of the amount of immobilized heparin on scaffolds with immersion time in PBS. The amount of immobilized heparin decreased from 41.78 ± 0.39 µg/scaffold to 22.33 ± 0.68 µg/scaffold after 14 day immersion in PBS. Toluidine blue method is a facile and qualitative way to verify and determine the presence of conjugated heparin based on the formation of a strong purple complex with the sulfated polysaccharide. Although about half of the originally immobilized heparin was released in the aqueous PBS solution due to cleavage of the amine bond between carboxylic acid groups in heparin and amino groups in gelatin or de-entrapment of gelatin, there was still a substantial amount of immobilized heparin on the scaffold and this remaining heparin could be revealed by toluidine blue staining (Fig.4(b)). In comparison, the plain Ca-P/PHBV scaffolds used for control could only absorb toluidine blue and displayed the light blue color.

Due to problems encountered when attempts were made for contact angle measurement using scaffolds, solvent-cast PHBV films, instead of porous scaffolds, were made and used for measuring contact angles before and after surface modification. The contact angle for PHBV films before surface modification was $93.61\pm1.85°$. It decreased to $75.12\pm3.79°$ after gelatin entrapment. The physical entrapment of gelatin improved the hydrophilicity of PHBV and hence it may be inferred that it can also improve the hydrophilicity of sintered scaffolds. After the immobilization of heparin, the contact angle further decreased, albeit slightly, to $72.06\pm0.89°$. Fig.5 shows the contact angle images of PHBV film before and after surface modification.

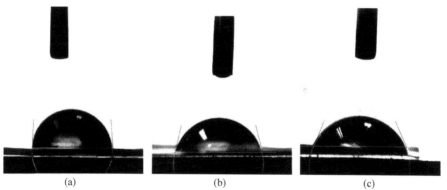

(a)   (b)   (c)

Figure 5. Images for the contact angle of PHBV film (a), PHBV entrapped with gelatin (b), and PHBV immobilized with heparin (c).

CONCLUSIONS

Three dimensional Ca-P/PHBV nanocomposite scaffolds were produced via SLS and gelatin was successfully entrapped in the surface of sintered scaffolds through physical entrapment. The amount of entrapped gelatin increased significantly with increasing TFE concentration and treatment time. It was shown that heparin could be immobilized on the scaffolds through the reaction of NHS-activated carboxylic acid groups of heparin with residual primary amino groups in gelatin-entrapped scaffolds. More than half of the heparin originally conjugated on the scaffolds still remained immobilized on scaffolds after 14 day immersion in PBS. After gelatin entrapment and subsequent heparin immobilization, the hydrophilicity of the PHBV matrix was greatly improved and the contact angle decreased from 93.61±1.85° to 72.06±0.89°. The surface modification improves the hydrophilicity of the scaffolds and the heparin immobilized on scaffolds will provide the binding site for growth factors, leading to their sustained release during bone tissue regeneration.

ACKNOWLEDGEMENTS

B.Duan thanks The University of Hong Kong (HKU) for making the University Scholarship Award to him. This work was supported by an HKU research grant and also partly by a CERG grant (HKU 7182/05E) from the Research Grants Council of Hong Kong. Assistance provided by technical staff in the Department of Mechanical Engineering, HKU, is acknowledged.

REFERENCES

[1]K.F. Leong, C.M. Cheah, and C.K. Chua, Solid freeform fabrication of three-dimensional scaffolds for engineering replacement tissues and organs, *Biomateirals*, **24**, 2363-78 (2004).

[2]W.Y. Yeong, C.K. Chua, K.F. Leong, and M. Chandrasekaran, Rapid prototyping in tissue engineering: challenges and potential, *Trends. Biotechnol.*, **22**, 643-52 (2004).

[3]A. J. Salgado, O.P. Coutinho, and R.L. Reis, Bone tissue engineering: State of the art and future trends, *Macromol. Biosci.*, **4**, 743-65 (2004).

[4]J.E. Babensee J, L.V. McIntire, and A.G. Mikos, Growth factor delivery for tissue engineering, *Pharm. Res.*, **17**, 497-504 (2000).

[5]I. Freeman, and S. Cohen, The influence of the sequential delivery of angiogenic factors from affinity-binding alginate scaffolds on vascularization, *Biomaterials*, **30**, 2122-31 (2009).

[6]J.S. Lee, J.W. Bae, Y.K. Joung, S.J. Lee, D.K. Han, and K.D. Park, Controlled dual release of basic fibroblast growth factor and indomethacin from heparin-conjugated polymeric micelle, *Int. J. Pham.*, **346**, 57-63 (2008).

[7]B. H. Zhao, T. Katagiri, H. Toyoda, T. Takada, T. Yanai, T. Fukuda, U.I. Chung, T. Koike, K. Takaoka, and R. Kamijo, Heparin potentiates the in vivo ectopic bone formation induced by bone morphogenetic protein-2, *J. Biol. Chem.*, **281**, 23246-53 (2006).

[8]O. Jeon, S.J. Song, S-W. Kang, A.J. Putnam, and B-S. Kim, Enhancement of ectopic bone formation by bone morphogenetic protein-2 released from a heparin-conjugated poly(L-lactic-co-glycolic acid) scaffold, *Biomaterials*, **28**, 2763-71 (2007).

[9]H. Lin, Y.N. Zhao, W.J. Sun, B. Chen, J. Zhang, W.X. Zhao, Z.F. Xiao, and J.W. Dai, The effect of crosslinking heparin to demineralized bone matrix on mechanical strength and specific binding to human bone morphogenetic protein-2, *Biomaterials*, **29**, 1189-97 (2008).

[10]S.Y. Meng, M. Rouabhia, G.X. Shi, and Z. Zhang Z, Heparin dopant increases the electrical stability, cell adhesion, and growth of conducting polypyrrole/poly(L,L-lactide) composites, *J. Biomed. Mater. Res. A*, **87A**, 332-44 (2008).

[11]D.H. Go, Y.K. Joung, S.Y. Park, Y.D. Park, and K.D. Park, Heparin-conjugated star-shaped PLA for improved biocompatibility, *J. Biomed. Mater. Res. A*, **86A**, 842-8 (2008).

[12]B. Duan, M. Wang, W.Y. Zhou, and W.L. Cheung, Totally Bioresorbable Tissue Engineering Scaffolds Based on Ca-P/PHBV Nanocomposite and Fabricated via Selective Laser Sintering: Part II. Scaffold Fabrication and Characterization. 8th World Biomaterials Congress. Amsterdam, 2008.

[13]B. Duan, M. Wang, W.Y. Zhou, and W.L. Cheung, Synthesis of Ca-P nanoparticles and fabrication of Ca-P/PHBV microspheres for bone tissue engineering applications, *Appl. Surf. Sci.*, **255**, 529-33

(2008).

[14]A. Kumar, V. Bansal, K.S. Nandakumar, I.Y. Galaev, P.K.R. Roychoudhury, R. Holmdahl, and B. Mattiasson, Integrated bioprocess for the production and isolation of urokinase from animal cell culture using supermacroporous cryogel matrices, *Biotechnol. Bioeng.*, **93**, 636-46 (2006).

[15]H.J. Chung, H.K. Kim, J.J. Yoon, and T.G. Park, Heparin immobilized porous PLGA microspheres for angiogenic growth factor delivery, *Pharm. Res.*, **23**, 1835-41 (2006).

[16]X.H. Liu, Y.J. Won, and P.X. Ma, Surface modification of interconnected porous scaffolds, *J. Biomed. Mater. Res. A*, **74A**, 84-91 (2005).

LAYERED DOUBLE HYDROXIDE: A NEW CERAMIC-BASED HEMOSTATIC AGENT?

Z. Tahmasebi-Birgani[1], M. Solati-Hashjin[1], H. Peirovi[2], S. Shafiei[1], A. Farzadi[1], A. Aminian[1]

[1]Amirkabir University of technology, Tehran, Iran.

[2]Nanomedicine & Tissue Engineering Research Center, Shahid Beheshti University of Medical Sciences, Tehran, Iran.

ABSTRACT

Layered double hydroxides (LDHs), well known as anionic clays or hydrotalcite-like compounds consist of positively charged layers. Net positive charges on the layers are balanced by exchangeable anions along with water molecules in the interlayer space. In the present research, the potential for use of Ca/Al-LDH as local hemostatic agent is evaluated in vitro by measuring the coagulation time of human fresh blood in the presence of LDH powder. Crystalline Ca/Al-LDH powders required for these experiments were synthesized via a co-precipitation method followed by a controlled hydrothermal process using calcium nitrate and aluminum nitrate as starting materials under certain synthetic conditions such as time and concentration. XRD and SEM analyses were utilized to characterize the synthetic powders. XRD and SEM results demonstrate the presence of a crystalline pure Ca/Al-LDH with hexagonal morphology. In order to investigate the biocompatibility of the samples, culture techniques were employed using mouse fibroblast cells (L929) as reference cell line. The results reveal that LDH powder can accelerate the platelet aggregation and the formation of blood clot. Moreover, the samples show good cell proliferation and high cell viability to L929 cell line compared with negative control sample. These results demonstrate that these layered materials are suitable candidates for using as local hemostatic agent to hemorrhage control and can reduce blood loss in lethal injuries.

INTRODUCTION

Uncontrolled bleeding and consequent hemorrhagic shock are leading causes of death in many injuries and early hemorrhage control can improve outcome [1-2]. In minor accidents, the bleeding may be controlled by the body's own blood clotting mechanisms. In more severe wounds, additional controls such as applying pressure or using of absorbent dressings or pads are required. However, these methods may be ineffective for severe bleedings or bleedings in persons with lessened blood clotting mechanisms. One efficient way to control bleeding is to use a hemostatic agent in the form of powder, foam, pad, etc. at the source of the bleeding. Limited materials are known for hemorrhage control [3]. Such an agent must be safe, rapidly effective and have good systemic and local compatibility [4]. Several fibrin-based, collagen-based, cellulose-based and gelatin-based [4-7] systems have been introduced as local hemostatic agents. While these materials are effective for control of bleeding, they are expensive.

A newer local hemostatic agent –commercially named QuikClot®- is a zeolite-based powder that shows good control in many bleedings with different bleeding sources [1-2, 8-9]. Zeolites are a family of Silicates that are one of the most abundant classes of minerals on Earth. Clays are a group of inorganic materials that have structural similarities to zeolites [10].

Layered Double Hydroxides (LDHs), also called anionic clays or hydrotalcite-like compounds, are clay-like materials that show promising properties for a large number of applications. Layered double hydroxides (LDHs) have been known for over 150 years since the discovery of the mineral hydrotalcite. They are a broad class of inorganic lamellar compounds with high capacity for anion intercalation.

The LDH structure -represented by $[M^{2+}_{1-x}M^{3+}_x(OH)_2]^{x+} (A^{n-})_{x/n} \cdot mH_2O$- results from the stacking of brucite-like layers ($[Mg(OH)_2]$) containing extra positive charge due to the partial isomorphous substitution of M^{2+} by M^{3+}. This positive excess charge is balanced by anions, which exist in the interlamellar spaces [1-3]. Layered double hydroxides display unique physical and chemical properties close to those of clay minerals. Some interesting properties of these materials are summarized as follows: high specific surface area, swelling property, memory effect, high anion exchange capacity, and dispersive property. These interesting properties along with low or null toxicity and good biocompatibility make LDHs suitable candidates for medical and biological applications [11-12].

Some of LDHs' applications include applying in catalysts, adsorbents, and polymer composites. The application of LDHs in medical diagnostic systems such as biosensing and medical imaging systems is also well-known [13].

One of the most interesting biological applications of LDH is in pharmaceutical area especially in drug and gene delivery systems. These applications rely on acid buffering effect and anion exchange property. LDHs can intercalate many important biomolecules with negative charge such as drugs, oligomers, single or double stranded DNA, and simple molecules like nucleotides. The possibility of layered double hydroxides as molecular containers for pharmaceutical agents has also to be considered [11-12].

In this study, the potential of Ca_2Al-LDH powder for using as a local hemostatic agent material is evaluated.

METHOD

Sample Preparation
Ca_2Al-LDH were synthesized via a simple co-precipitation method using calcium and aluminium nitrate salts as starting materials followed by a hydrothermal post-treatment. The clear metal solutions were prepared with the total metal concentration 0.9 M. The ratio of Mg/Al was fixed at 2/1. The clear solutions were then mixed. The solution of metal nitrates was titrated up to pH=11 with 2M of NaOH solution at room temperature under Nitrogen atmosphere and then centrifuged. The obtained precipitates were aged in autoclave at 100° C for 1 hour and then freeze-dried for 24 hours.

Characterization
The powder X-ray diffraction patterns for the precipitates were obtained by means of a Siemens D5000 diffractometer with CuKa radiation (k = 1.5418 A °). The size and morphology of the product particles were measured using Philips XL30 scanning electron microscope.

Biological Experiments
The potential for use of Ca_2Al-LDH powder as hemostatic agent material was investigated in vitro by measuring and comparing the clotting time (CT) of fresh human blood plasma in the presence certain amount of LDH powder and comparing it with normal CT.

In order to investigate the biocompatibility of LDH, cell culture techniques were employed using mouse fibroblast cells (L929) as reference cell line. The cells were incubated in RPMI-1640 culture medium in a 12-well culture plate the presence of samples in humid atmosphere with 5% CO_2 at 37° C for a week. 1 well including culture medium and cell suspension was considered as negative control.

RESULT AND DISCUSSIONS
According to the X-ray diffraction pattern as shown in figure 1, layered double hydroxide powder is well crystallized and exhibits sharp and symmetric peaks at low 2θ angles in the one hand, but broad and asymmetric ones at higher 2θ angles, which are characteristics of clay minerals with a layered structure [14-16]. No impurity or second phase was observed in the XRD pattern of layered double hydroxide.

As can be seen in SEM micrographs (Figure 2), LDH powder has well formed and regular hexagonal shaped particles stacked on top of each other.

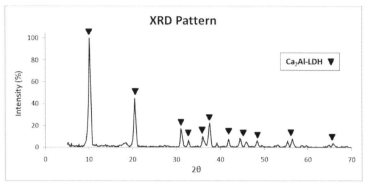

Figure 1: XRD pattern of LDH powder

The results of CT tests are shown in Table 1. As shown in this Table, the use of Ca_2Al-LDH as a blood coagulation accelerator can reduce the clotting time up to 10% in compare with control sample. This may be due to the absorption of a certain amount of blood water by layered structure or the attraction of platelets by positively charged layers. Although this reduction comparing those caused by the presence of other hemostatic agent such as QuikClot [1-2], is slight, but structural or physical modifications can improve its coagulation effects. These modifications include using other divalent and trivalent cations such as Mg^{2+}, increasing the surface area of the LDH powder, using of dehydrated LDH, etc.

As can be seen in figure 3., L929 cells can attach to and spread on the surface of the LDH sample similar to negative control (polystyrene plate). This result shows that LDH has no cytotoxicity for this cell line.

Figure 2: SEM images of LDH powder

Table 1. The results of CT test

Sample	The mass of powder (mg)	Blood volume (CC)	CT (S)
Control	0	1	292
Ca_2Al-LDH	20	1	250

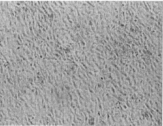

Figure 3: The results of cell culture test (left: LDH sample and right: negative control)

CONCLUSION

Layered double hydroxide powder was synthesized successfully using a simple co-precipitation method followed by a hydrothermal post-treatment. The results of XRD and SEM analyses revealed the presence of high crystalline pure LDH with hexagonal morphology. As described in the results, LDH powder reduced the blood coagulation time but this reduction is very slight. The determination of the precise mechanism of this phenomenon and improve the coagulation characteristic of LDH require more studies.

REFERENCES

[1]H. B. Alam, G.B. Uy, D. Miller, E. Koustova, T. Hancock, R. Inocencio, D. Anderson, O.Llorente, and P. Rhee, Comparative Analysis of Hemostatic Agents in a Swine Model of Lethal Groin Injury, J. TRAUMA., 54, 1077-82 (2003).

[2]H.B. Alam, Z. Chen, A. Jaskille, R.C. Querol, E. Koustova, R. Inocencio, R. Conran, A. Seufert, N. Ariaban, K. Toruno, and P. Rhee, Application of a Zeolite Hemostatic Agent Achieves 100% Survival in a Lethal Model of Complex Groin Injury in Swine, J. TRAUMA., 56, 974-83 (2004).

[3]Internatinal Application Published Under The Patent Cooperetaion Treaty(PCT), International Publication Number WO 02/30479 A1.

[4]L.K. Krishnan, M. Mohanty, P.R. Umashankar, and A.V. Lal, Comparative evaluation of absorbable hemostats: advantages of fibrin-based sheets, Biomaterials, 25, 5557–63 (2004).

[5]G. Dickneite, H. Metzner, T. Pfeifer, M. Kroez, and G. Witzke, A comparison of fibrin sealants in relation to their in vitro and in vivo properties, Thrombosis Research, 112, 73– 82 (2003).

[6]K.J. Smith, H.G. Skelton, T.L. Barrett, M. Welch, and J. Beard, Histologic and immunohistochemical features in biopsy sites in which bovine collagen matrix was used for hemostasis, J. AM. ACAD. DERMATOL., 34, 434-8 (1996).

[7]Y.M. Hong, and K.R. Loughlin, The Use of Hemostatic Agents and Sealants in Urology, J. UROLOGY., 176, 2367-74 (2006).

[8]V. Margulis, E.D. Matsumoto, R. Svatek, WA. Kabbani, J.A. Cadeddu, and Y. Lotan, Application of Novel Hemostatic Agent During Laparoscopic Partial Nephrectomy, J. UROLOGY., 174, 761–4 (2005).

[9]A. Mahajna, M.H. Michael, and M. Krausz, Use of the Hemostatic Agent QuikClot® for the Treatment of Massive Splenic Injury in a Rat Model, Eur. Surg. Res., 39, 251–7 (2007).

[10]P. Boulet, H.C. Greenwell, S. Stackhouse, and P.V. Coveney, Recent advances in understanding the structure and reactivity of clays using electronic structure calculations, J. Molecular Structure: THEOCHEM, 762, 33–48 (2006).

[11]C.D. Hoyo, Layered double hydroxides and human health: An overview, App. Clay Sci., 36,103–21 (2007).

[12] J. Choy, S. Choi, J. Oh and T. Park, Clay minerals and layered double hydroxides for novel biological applications, App. Clay Sci., **36**, 122–32 (2007).

[13] S. Shafiei, Z. Tahmasebi-Birgani, A. Darvish, M. Azimi, and M. Solati-Hashjin, Layered Double Hydroxides for Diagnostic Applications, Internatinal Congress of Evalution of Medical Diagnosis Modern Technologies, (2008).

[14] J.M. Oh, S.H. Hwang, and J.H. Choy, The effect of synthetic conditions on tailoring the size of hydrotalcite particles, Solid State Ionics, **151**, 285–91(2002).

[15] G.R. Williams, and D. O'Hare, New phosphonate intercalates of [Ca2Al(OH)6]NO3·yH2O: A synthetic and kinetic study, Solid State Sci., **8**, 971–80 (2006).

[16] L. Raki, J. Beaudoin, and L. Mitchell, Layered double hydroxide-like materials: nanocomposites for use in concrete, NRCC, **45219**, 27-8 (2004).

MICROWAVE-ASSISTED SYNTHESIS AND CHARACTERIZATION OF BIPHASIC CALCIUM PHOSPHATE NANOPOWDERS

A. Farzadi, M. Solati-Hashjin, Z. Tahmasebi-Birgani, A. Aminian
Nanobiomaterials Laboratory, Biomaterial Group, Faculty of Biomedical Engineering, Amirkabir University of Technology
Tehran, Tehran, Islamic Republic of Iran

ABSTRACT
    The most-widely used calcium phosphate based bioceramics are hydroxyapatite (HA) and beta-tricalcium phosphate (β-TCP). Biphasic calcium phosphate (BCP) ceramics are ideal bone substitutes since their bioactivity and biodegradation can be adjusted by varying HA/β-TCP ratio. BCP powders of β-TCP and HA were synthesized by microwave assisted acid-base reaction. The powders obtained by adding dropwise orthophosphoric acid to the suspension of calcium hydroxide. Initial Ca/P ratio was adjusted to 1.64 or 1.53 in order to obtain mixtures of HA and β-TCP in different ratios in the resulting BCP powders. Sodium hydroxide or nitric acid was used to control pH. The pH was 12 at the initial stage of the reaction but dropped to 7 and 9 in the synthesis of HA-riched BCP and β-TCP-riched BCP respectively. The resulting precipitate was calcined in air at 900° C for one hour. The phase analysis of BCP nanopowders was identified using an X-ray diffractometer. The scanning electron microscopy technique was employed to study the final microstructure of the synthesis powders. The results show that the spherical nanoparticles of biphasic calcium phosphate can be successfully synthesized under optimized conditions. Microwave-assisted synthesis offers the advantage of efficient transformation of energy in shorter time period which leads to finer crystalline structure.

INTRODUCTION
    During the past 50 years, research and development in specific bioceramics such as alumina, zircon, calcium phosphates and bioglass have made significant contributions to healthcare and quality of human life. These bioceramics can be used in the human body to substitute for damaged segments of skeleton system[1-3]. The most widely used calcium phosphates are hydroxyapatite and beta tricalcium phosphate. These biomaterials are well known bioceramics that can provide an interfacial bond between an implant and surrounding tissues. This is due to similarity of such materials to the mineral constituents of human bone and teeth[3-5]. HA is stable in the body fluid while TCP is rather soluble, so the dissolution rate of HA in body fluid is too low. On the other hand, the dissolution rate of β-TCP is too fast for bone bonding[4]. Therefore biphasic calcium phosphate consisting of HA and TCP can be used to control the bioresorbability and achieve optimal results[5,6]. Current methods for production of BCP are not satisfactory and suffer from a number of drawbacks such as: long synthesis time, low quality control and chemical contamination[7]. HA can be synthesized by most of the methods but pure TCP is usually produced in a solid state reaction. The microwave synthesis methods have been reported for the preparation of HA and TCP within a shorter period of time[6-8]. This method offers the advantages of heating throughout the volume and efficient transformation of energy[9].

EXPERIMENTAL
Materials
    Calcium hydroxide (Merck 1.02047.0500) and Orthophosphoric acid (Merck 1.00563.2500) were used as the sources of Ca^{2+} ion and PO_4^{3-} ion respectively. Nitric acid and sodium hydroxide were used to adjust the pH of the emulsion system.
Synthesis of nanopowders

Two different synthesis methods were used for preparation of the BCP powders: (1) Nanopowders were produced in a wet chemical, acid-base reaction. (2) Samples were prepared by chemical reaction using microwave radiation. The phase transformation and morphology of the resulting powders were characterized using X-Ray Diffraction (XRD), Fourier Transform Infrared Spectroscopy (FTIR), and Scanning Electron Microscopy (SEM) techniques. The Simulated Body Fluid (SBF) analysis was performed for the characterization of nanopowders depending on the pH of the starting solution and the presence of TCP in the powders.

(1) Preparation of biphasic calcium phosphate by acid-base reaction

Calcium hydroxide powder was suspended in 250ml of distilled water at room temperature and the powders were obtained by adding dropwise H_3PO_4 solution to the suspension. The suspension was centrifuged, dried out in an oven at 90° C overnight and calcined at 900° C for 1 hour. The initial Ca/P ratio, pH of the suspension and the aging time are given in Table I.

(2) Preparation of biphasic calcium phosphate by microwave assisted synthesis

The conditions for the preparation of BCP powders using microwave irradiation were the same as before, but after preparation of the solution mixture, it was immediately transferred to a microwave oven (2.45 GHz, 800W) and was irradiated for 45 min. The irradiation was carried out until the complete formation of the precipitate. The precipitate was then centrifuged, dried out and calcined. The conditions of synthesis used for BCP preparation are given in Table II.

Table I: Synthesis conditions for BCP preparation for method 1

Sample	HA-Riched-BCP	TCP-Riched-BCP
	BCP1	BCP2
Initial Ca/P ratio	1.54	1.53
pH	Basic	Acidic
Aging time	10 hrs	6 hrs

Table II: Synthesis conditions for BCP preparation for method 2

Sample	HA-Riched- BCP (MW)	TCP-Riched-BCP (MW)
	BCP3	BCP4
Initial Ca/P ratio	1.54	1.53
pH	Basic	Acidic
Aging time	45 min	45 min

CHARACTERIZATION

The phase transformation, chemical components, morphology and particle size were characterized using XRD, FTIR and SEM methods respectively. The samples were put in SBF to investigate the biodegradable behavior of BCP powders consisting of HA and TCP. The characterization results were compared to show the effect of microwave presence in synthesis of calcium phosphate powders.

RESULTS AND DISCUSSION

The XRD patterns of BCP powders are shown in Figure 1. The XRD patterns of powders prepared by microwave radiation are similar to those prepared without it. It is clear that using microwave increases the peak heights which correspond to the increase in crystallinity[10].

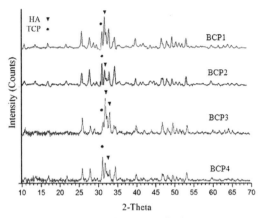

Figure 1: The XRD patterns of BCPs powders

The FTIR spectra of prepared powders are shown in Figure 2. These spectra illustrate the bonds of OH^- at 3550 cm^{-1} and HOP_4^{-2} at 970cm^{-1} related to HA and β-TCP structure respectively. The powders prepared by microwave assisted method have the FTIR patterns similar to Figure 2.a[11,12].

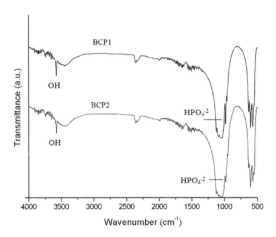

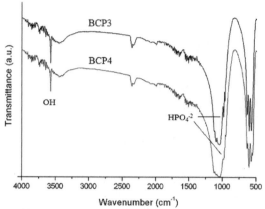

Figure 2 (a,b): The FTIR spectra of BCPs powders

Figure 3 shows the percentages of HA phase in the powders after using microwave. The behavior of dipole moment of the hydroxyl ions in HA structure could be responsible for the dielectric nature of hydroxyapatite. OH⁻ groups absorbed more microwave radiation which indicates the development of HA phase[13]. The XRD and FTIR patterns illustrate this result too.

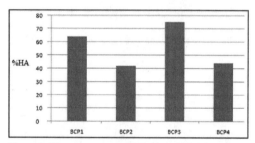

Figure 3: The percentages of HA phase in BCPs powders

In Figure 4 the morphologies of the powders are shown. These particles have spherical shape in nanometric scale, but samples produced with microwave become fine and more homogeneous. The SBF solution was used to characterize the biodegradable behavior of samples under a controlled environment of 37° C and pH 7.25[14]. The powders used as sintered pellet shape. The pH values were recorded at regular time intervals that are shown in Figure 5.

Figure 4: The morphology of BCPs powders (×15000)

The BCP1 sample shows a decrease in the pH value but less than that of BCP2. This is due to more tricalcium phosphate phase presence in the BCP2 sample that is converted to a stable HA phase causing the acidic ion of HPO_4^{-2} to be released. Also, the sample prepared by microwave assisted method showed less decrease in the pH value indicating its higher stability because of increase in crystallinity[8,15].

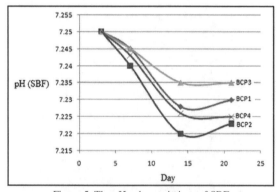

Figure 5: The pH value variations of SBF

CONCLUSION

Microwave assisted synthesis of biphasic calcium phosphate powders can: (a) decrease synthesis time from hours to just 45 minutes, (b) improve crystallinity, (c) increase the amount of HA phase. HA/β-TCP ratio is a function of Ca/P ratio and pH of the suspension. The dissolution of BCP powder is strongly dependent on the β-TCP content and the crystallinity of the powder. In the future, it may be possible to control the bioresorbability of the calcium phosphate powders by adjusting the β-TCP percentage.

ACKNOWLEDGEMENTS
The authors gratefully acknowledge to Dr. A. Mahyari and S. Emami for their valuable help in the preparation of the manuscript.

REFERENCES
[1]S. J. Kalita, A. Bhardwaj and H. A. Bhatt, Nanocrystalline calcium phosphate ceramics in biomedical engineering, *Materials Science and Engineering C*, **27**, 441–449(2007)
[2]T. V. Thamaraiselvi and S. Rajeswari, Biological Evaluation of Bioceramic Materials - A Review, *Trends Biomater. Artif. Organs*, **18**, 9-17(2004).
[3]M. Solati-Hashjin, Effect of powder synthesis route on microstructure and of properties of hydroxyapatite, *PhD Thesis (Materials Engineering), Materials and Energy Research Center, Tehran, Iran*, 1998.
[4]Y. Li, W. Weng and K. Ch. Tam, Novel highly biodegradable biphasic tricalcium phosphates composed of α-tricalcium phosphate and β-tricalcium phosphate, *Acta Biomaterialia*, **3**, 251–254 (2007).
[5]A. Farzadi, Microwave-assisted synthesis and characterization of biphasic calcium phosphate nanopowders, *MSc Thesis (Biomedical Engineering), Amirkabir University of Technology, Tehran, Iran*, 2009.
[6]M. Vallet-Reg, J. Mar and G. Calbet, Calcium phosphates as substitution of bone tissue, *Progress in Solid State Chemistry*, **32**, 1-31(2004).
[7]M. Youn, R. Kanti Paul, H. Song and B. Lee, Fabrication of Porous Structure of BCP Sintered Bodies Using Microwave Assisted Synthesized HAp Nano Powder, *Materials Science Forum*, **534-536**, 49-52 (2007).
[8]I. Manjubala and M. Sivakumar, In-situ synthesis of biphasic calcium phosphate Ceramics using microwave irradiation, *Materials Chemistry and Physics*, **71**, 272–278 (2001).
[9]B. Lee, M. Youn, R. Paul, K. Lee and H. Song, In situ synthesis of spherical BCP nanopowders by microwave assisted process, *Materials Chemistry and Physics*, **104**, 249–253 (2007).
[10]Joint Committee on Powder Diffraction Standards, Powder Diffraction Files 09-0432 and 09-0169.
[11]H. Zhang, Y. Yan, Y. Wang and S. Li, Morphology and Formation Mechanism of Hydroxyapatite Whiskers from Moderately Acid Solution, *Materials Research*, **6**, 111-115 (2002).
[12]A. Siddharthan, S. K. Seshadri and T. S. Sampath Kumar, Rapid Synthesis of Calcium Deficient Hydroxyapatite Nanoparticles by Microwave Irradiation, *Trends Biomater. Artif. Organs*, **18**, (2005).
[13]M. P. Mahabole, R. C. Aiyer, C. V. Ramakrishna, B. Sreedhar and R. S. Khairnar, Synthesis, characterization and gas sensing property of hydroxyapatite ceramic, *Bull. Mater. Sci.*, **28**, 535–545 (2005).
[14]A. Jaroenworaluck, N. Koolpreechanan, N. Kosachan and R. Stevens, Surface Modification of HA and TCP immersed in Simulated Body Fluid, *European Cells and Materials*, **14**, Suppl. 1 (2007).
[15]S. Kwon, Y. Jun, S. Hong and H. Kim, Synthesis and dissolution behavior of b-TCP and HA/b-TCP composite powders, *Journal of the European Ceramic Society*, **23**, 1039–1045 (2003).

SILICON-SUBSTITUTED HYDROXYAPATITE SYNTHESIZED BY A HYDROTHERMAL METHOD

A. Aminian, M. Solati-Hashjin, F. Bakhshi, A. Farzadi
Nanobiomaterials Lab, Biomaterials Group, Faculty of Biomedical Engineering, Amirkabir University of Technology
Tehran, I. R. Iran

ABSTRACT

Silicon-substituted hydroxyapatite (Si-HA) bioceramics are widely used as bone replacement materials. The incorporation of silicone into hydroxyapatite (HA) has been shown to increase the rate of bone apposition into HA bioceramics implants significantly. Si, an essential trace element required for healthy bone and connective tissues, influences the biological activity of CaP (calcium phosphate) materials by modifying material properties and it has direct effects on the physiological processes in skeletal tissue. Silicon has been substituted in the HA lattice by replacing some of the phosphate (PO_4) groups by silicate (SiO_4) groups resulting in silicate substituted HA described as $Ca_{10}(PO_4)_{6-x}(SiO_4)_x(OH)_{2-x}$. Si-HA was prepared successfully by hydrothermal method, using $Ca(NO_3)_2$, $(NH_4)_3PO_4$ and $Si(OCH_2CH_3)_4$ TEOS as starting materials. The analysis of samples by ICP, XRD, FTIR and SEM showed that the substitution of phosphate groups by silicate groups causes some OH⁻ loss to maintain the charge balance and changes of the lattice parameters were observed. Furthermore the substitution of the silicate groups reduces the size of Si-HA crystallites.

Keywords: Hydroxyapatite; Silicon; Silicon-Substituted Hydroxyapatite; Bioactivity; Non-Stoichiometric;

INTRODUCTION

Since bone mineral has a similar crystallographic structure to hydroxyapatite (HA), HA has achieved significant application as a bone graft material in a wide range of medical and dental applications [1-4].

The chemical formula of stoichiometric HA is $Ca_{10}(PO_4)_6(OH)_2$ with a Ca:P ratio of 5:3 (1.67). HA present in hard tissue, however, exhibits calcium, phosphate and hydroxyl deficiency (reported Ca:P ratios 1.37-1.87 [5]), internal crystal disorder and contains various cationic and anionic substituents [2,6]. The type and the amount of ionic substitution in biological apatite varies from 2 to 8 wt.% in CO_3, 1 wt.% in Mg, and 0.4 to 0.8 wt.% for Si. Although these substitution rates are low, it is well understood that these elements are associated with the properties of biological apatites and play a major role in the biochemistry of bone, enamel and dentin [7-9].

On the other hand, bioactivity of HA implants in comparison to bioglass and glass ceramic materials, is low [4,7]. A possible approach to improve the bioactivity of HA ceramics is to incorporate small amounts of physiologically relevant ions into the HA lattice. It has been shown that silicon-substituted hydroxyapatite ceramics, exhibit enhanced in vitro and in vivo bioactivity compared to phase pure HA [3,8,10-15]. Several authors have worked on this subject, assuming that Si or SiO_4^{4-} replaces P or PO_4^{3-}, which causes some OH⁻ loss to maintain the charge balance [13].

These ionic substitutions can affect the surface structure and charge of hydroxyapatite, which would improve the bioactivity of apatite. The presence of silicon in HA has shown to play an important role on the formation of bone [13,14].

Hench reported that the deterioration in the proliferation and function of osteoblasts due to osteopenia and osteoporosis is related to the loss of biologically available silicon [16], and Keeting reported that bone cells in culture proliferate more rapidly in the presence of soluble silicon [17]. Reffit demonstrated that physiological concentrations of soluble silicon stimulate collagen type I synthesis in human osteoblast-like cells and promote osteoblastic differentiation [18]. Gibson also demonstrated that Si-HA increases the metabolic activity of human osteosarcoma cells. In vitro studies by Gibson et al. showed that the substitution of silicate ions for phosphate ions into hydroxyapatite enhances osteoblast cell activity, compared to phase pure HA [19].

These studies clearly demonstrate the possible advantages of incorporation of silicon into the lattice of biomaterials intended for applications in bone implants. The aim of this study is to synthesize Si-substituted hydroxyapatite by hydrothermal method to improve bioactivity of hydroxyapatite.

MATERIALS AND METHODS

Stoichiometric hydroxyapatite and Si-substituted hydroxyapatite were prepared by hydrothermal method using $Ca(NO_3)_2.4H_2O$ (Prolabo Merck Eurolab NO.22384.298) as a source of Ca, $(NH_4)_3PO_4$ (Ridel − de Haen No.05447) as a source of P and $Si(OCH_2CH_3)_4$: (TEOS) (Merck NO.8.00658) as a source of Si. The amount of reagents (Table 1) was calculated on the assumption that silicon would substitute phosphorus.

Table 1. Quantities of reactants used and the measured wt.% of Si of the samples

Samples	$n(Ca(NO_3)_2)$/mol	$n((NH_4)PO_4)$/mol	$n(TEOS)$/mol ($\times 10^{-4}$)
HA T0.0SC*	0.025	0.0150	0.00
0.8 wt.%Si-HA T0.8SC*	0.025	0.0143	0.71
1.5 wt.%Si-HA T1.5SC*	0.025	0.0132	1.34
1.5 wt.%Si-HA T1.5SR**	0.025	0.0132	1.34

*C: Calcinated
**R: Not calcinated

The $Ca(NO_3)_2$ solution (0.5 M) was prepared keeping the pH higher than 10.0. Simultaneously the $(NH_4)_3PO_4$ solution (0.25 M) was prepared and its pH was kept higher than 11.0 by the addition of NH_3 solution (Merck NO.1.05426). $Ca(NO_3)_2.4H_2O$ solution containing 0.2 g polyethylene glycol was added drop-wise to $(NH_4)_3PO_4$ and TEOS solutions, and the reaction mixture was stirred for 0.5 h followed by hydrothermal treatment at 200°C for 8 h. The resulting precipitates were centrifuged three times, and then dried at 100°C for 12 h. A fraction of each as-prepared sample was treated at 800°C for 1 h under air atmosphere, respectively.

The chemical composition (Ca, P, Si contents) was determined by inductively coupled plasma (ICP) using an ICP − AES ARL− 3410 spectrometer. The phase compositions of powders were determined using X-ray diffraction (XRD). The XRD data were collected over 2θ range of 10-60° with a step size of 0.02° by Siemens D500 diffractometer (40 KV and 30 mA) using Cu-Kα radiation (1.54Å). Phase identification was achieved by comparing the diffraction patterns of HA and Si-HA with ICDD (JCPDS) standards. The average crystallite size of the samples was calculated by using the Scherrer formula. Compositions were also studied by Fourier transform infrared (FTIR) spectroscopy (Vector 33), in transmission mode using KBr pellets. A Philips XL30 scanning electron microscopy (SEM) was employed for microstructural characterization of HA and Si-HA samples.

RESULTS AND DISCUSSION

The chemical compositions of the Si-substituted samples as determined by ICP are shown in Table 2. together with the calculated Ca/P and Ca/(P+Si) ratios. The silicon content is lower than that of the corresponding amount of starting material, which implies that some of silicon ions remain in the mother liquor solution after precipitation. Because of the presence of CO_2 in solution during the synthesis, carbonate groups can replace some phosphorus sites and hence carbonated apatite was formed. Therefore the amount of phosphorus substitution by silicon was reduced [20].

Table 2. Chemical analysis of the samples

Samples	(wt.%)			(Weight Ratio)		(Atomic Ratio)	
	Ca	P	Si	Ca/P	Ca/(P+Si)	Ca/P	Ca/(P+Si)
HA (T0.0SC)	39.88	18.49	0.00	2.157	2.157	1.667	1.667
0.8 wt.% Si-HA (T0.8SC)	—	—	—	—	—	—	—
1.5 wt.% Si-HA (T1.5SC)	39.75	17.19	1.35	2.312	2.144	1.787	1.657

X-ray diffraction patterns of powders are shown in Fig. 1. The X-Ray diffraction patterns of samples calcined at 800°C for 1 h indicate that all the peaks in the diffraction pattern can be indexed on the basis of JCPDS Card No.09-0432 and do not reveal the presence of extra phases related to silicon oxide species or other calcium phosphate species. XRD analysis of the Si-HA showed a single, crystalline phase, similar to stoichiometric HA. Silicon substitution obviously does not affect the diffraction pattern of hydroxyapatite.

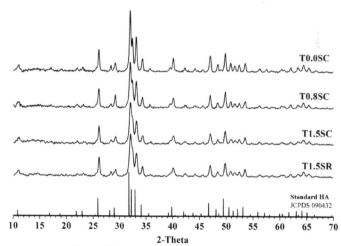

Fig .1. XRD patterns of HA and Si-HA samples

Nevertheless, all patterns of samples present a crystalline structure with exactly the same reflection of pure HA, but the corresponding HA reflections loose intensity with the increasing Si content, proving a progressive loss of crystallinity [12].

Lattice parameters of pure and Si-substituted Hydroxyapatite samples were determined by Rietveld structure refinement of X-ray diffraction data of each sample sintered at 800°C for 1 h [13, 21]. Unit cell parameters and unit cell volume of the samples are listed in Table 3.

Table 3. Lattice parameters of HA and Si-HA

Samples	a (Å)*	c (Å)*	v (Å^3)
HA (T0.0SC)	9.4082	6.8735	526.8921
0.8 wt.% Si-HA (T0.8SC)	9.4065	6.8740	526.74
1.5 wt.% Si-HA (T1.5SC)	9.4056	6.8790	527.0223
1.5 wt.% Si-HA (T1.5SR)	9.4060	6.8794	527.113

*Uncertainty: ±0.00005

Data of Tab. 3 show that silicon substitution results in a decrease in the a-axis and an increase in the c-axis of the unit cell of hydroxyapatite [7,13,20,22].

The average crystallite size of the samples was calculated using the Scherrer formula (1) [23]:

$$D_{hkl} = \frac{k\lambda}{Cos\theta\sqrt{\omega^2 - \omega_0^2}} \tag{1}$$

Where $D_{(hkl)}$ is the crystallite size (nm); k is the shape coefficient, 0.9; λ is the wave length (nm); θ is the diffraction angle(°); ω corresponds to experimental full width at half maximum (FWHM) obtained for each sample; ω_0 corresponds to standard FWHM [13,23]. Crystallite sizes calculated via the Scherrer formula are shown in Table 4.

Table 4. Calculated average crystallite size of HA and Si-HA via the Scherrer formula

Samples	T0.0SC	T0.8SC	T1.5SC	T1.5SR
Crystallite size (nm)	50.988	50.975	35.462	35.002

The crystallinity degree, corresponding to the fraction of crystalline phase present in the examined volume can be estimated by X-ray diffraction data and $X_C \approx 1 - \left(V_{112/300} / I_{300}\right)$ [21, 22], Where X_C is the crystallinity degree, I_{300} is the intensity of (300) reflection and $V_{112/300}$ is the intensity of the hollow between (112) and (300) reflection. The crystallinity degrees of different samples are listed in table 5.

Table 5. Estimated crystallinity degree of samples

Samples	T0.0SC	T0.8SC	T1.5SC	T1.5SR
Crystallinity degree (%)	80.75	79.05	68.80	50.26

Data of Tab. 5 show that silicon incorporation increases structural disorder (or decreases crystallinity) respectively.

FTIR was used to study the powders in the as-prepared state and in the heat-treated state to quantify the effect of the silicon substitution on the different functional groups, such as hydroxyl and phosphate groups of hydroxyapatite [13]. All FTIR spectra (Fig. 2) exhibited the characteristic OH^- band at 3570 cm^{-1} and PO_4^{3-} bands between 960 and 1100 cm^{-1} (1100, 1090, 1030 and 960 cm^{-1}) and 450 and 660 cm^{-1} (630, 600, 570 and 470 cm^{-1}) associated with HA. Broad bands due to adsorbed water (3000-2850 cm^{-1}) were also present, as were weak bands in the region associated with the CO_3^{2-}, vibration mode (1550-1410 cm^{-1}). The present absorptions indicate that the CO_3^{2-} group has substituted for both PO_4^{3-} and OH- groups in the HA structure [2,3,8,12,13,20,25]. Additionally two low intensity shoulders at 490 and 890 cm^{-1}, that did not appear in HA sample, can be observed as specific feature in Si-HA. These two peaks have been related to the presence of SiO_4^{4-} groups in the apatite structure [2,13,21,25,26].

Substitution of silicon into the hydroxyapatite $(Ca_{10}(PO_4)_{6-x}(SiO_4)_x(OH)_{2-x})$ reduces the amount of hydroxyl group to compensate for the extra negative charge of the silicate group. As the Si content is increased, there is a strong intensity loss of the CO_3^{2-} and OH^- related absorption bands [2,12,13,15].

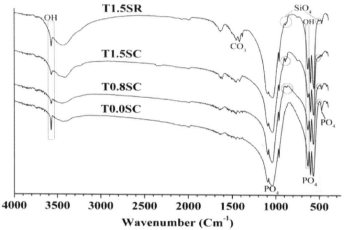

Fig .2. FTIR patterns of HA and Si-HA samples

SEM images of the specimens (Figs. 3-5) represent the morphology of HA and Si-HA. SEM results show a significant decrease in particle size in the Si-HA compared to pure HA[27], but there is no evidence that silicon incorporation into hydroxyapatite does affect the particle shape [28]. Particle sizes are in good accordance with crystallite size obtained by XRD data. Thus, it can be assumed that particles are single crystal rather than crystal agglomerates. Smaller crystal sizes of Si-bearing materials can be explained in terms of higher nucleation density during hydrothermal precipitation process.

Fig. 3. SEM Image of T0.0SC (Pure HA)

Fig. 4. SEM Image of T0.8SC (0.8% Si-HA)

Fig. 5. SEM Image of T1.5SC (1.5% Si-HA)

CONCLUSIONS

Silicon-substituted hydroxyapatite powders were synthesized successfully by hydrothermal method using $Ca(NO_3)_2$, $(NH_4)_3PO_4$ and $Si(OCH_2CH_3)_4$ (TEOS) as reagents. XRD analysis of the Si-HA showed a single, crystalline phase, similar to stoichiometric HA with no extra phase. The presence of Si decreases the crystallinity and increases the solubility of the powder, as substituting CO_3 does [7]. The analysis of FTIR shows that the substitution of phosphate groups by silicate groups causes some OH^- loss to maintain the charge balance and changes the lattice parameters of HA. Furthermore, SEM results show that incorporation of silicate groups reduces Si-HA particle size.

ACKNOWLEDGEMENTS
The authors gratefully acknowledge L. Hoseinzade and M. Kabiri for their valuable help in the preparation of manuscript and also Dr. Martin Schmücker for reviewing this paper.

REFERENCES
[1]L.C. Chow, Calcium Phosphate Materials: Reactor Response, *Adv Dent Res*, **2(1)**, 181-184 (1988).
[2]K. A. Hing, P. A. Revell, N. Smith, T. Buckland, Effect of Silicon Level on Rate, Quality and Progression of Bone Healin Within Silicate-substituted Porous Hydroxyapatite Scaffolds, *Biomaterials*, **27**, 5014–5026 (2006).
[3]M. Aizawa, N. Patel, A. E. Porter, S. M. Best, W. Bonfield, Syntheses of Silicon-containing Apatite Fibres by a Homogeneous Precipitation Method and Their Characterization, *Key Engineering Materials*, **309-311**, 1129-1132 (2006).
[4]A.E. Porter, N. Patel, J.N. Skepper, S.M. Best, W. Bonfield, Comparison of in vivo dissolution processes in hydroxyapatite and silicon-substituted hydroxyapatite bioceramics, *Biomaterials*, **24**, 4609–4620 (2003).
[5]A.S. Posner, Crystal chemistry of bone mineral, Physiol Rev, **49(4)**, 760–92 (1969).
[6]M. Solati Hashjin, F. Bakhshi, Biomimetic Synthesis of Nanostructured Magnesium Hydroxyapatite, *International Journal of Science & Technology Amirkabir*, **19**, No. 69-A, 47-51 (2009).
[7]E. Landi, S. Sprio, M. Sandri, A. Tampieri, L.Bertinetti, G. Martra, Development of Multisubstituted Apatites for Bone Reconstruction, *Key Engineering Materials*, **361-363**, 171-174 (2008).
[8]P.A.A.P. Marques, M.C.F. Magalhaes, R.N. Correia, synthesis and characterization of silicon-substituted hydroxyapatite, *Key Engineering Materials*, **192-195**, 247-250 (2001).
[9]A. M. Pietak, J. W. Reid, M. J. Stott, M. Sayer, Silicon substitution in the calcium phosphate bioceramics, *Biomaterials*, **28**, 4023–4032 (2007).
[10]N. Patel, E.L. Follon, I.R. Gibson, S.M. Best, W. Bonfield, Comparison of Sintering and Mechanical Properties of Hydroxyapatite and Silicon-Substituted Hydroxyapatite, *Key Engineering Materials*, *240-242*, 919-922 (2003).
[11]C.M. Botelho, R. A. Brooks, T. Kawai, S. Ogata, C. Ohtsuki, S.M. Best, M. A. Lopes, J.D. Santos, N. Rushton, W. Bonfield, In Vitro Analysis of Protein Adhesion to Phase Pure Hydroxyapatite and Silicon Substituted Hydroxyapatite, *Key Engineering Materials*, **284-286**, 461-464 (2005).
[12]E. L. Solla, F. Malz, P. González, J. Serra, C. Jaeger, B. León, The Role of Si Substitution into Hydroxyapatite Coatings, *Key Engineering Materials*, **361-363**, 175-178 (2008).
[13]X. L. Tang, X. F. Xiao, R. F. Liu, Structural characterization of silicon-substituted hydroxyapatite synthesized by a hydrothermal method, Materials Letters, **59**, 3841 – 3846 (2005).
[14]J. Lee, K. Lee, J. Chang, W. S. Cho, Y. Kim, S. Kim, Y. Kim, Biocompatibility of Si-Substituted Hydroxyapatite, *Key Engineering Materials*, **254-256**, 135-138 (2004).
[15]S.R. Kim, J.H. Lee, Y.T. Kim, D.H. Riu, S.J. Jung, Y.J. Lee, S.C. Chung, Y.H. Kim, Synthesis of Si,Mg substituted hydroxyapatites and their sintering behaviors, *Biomaterials*, **24**, 1389–1398 (2003).
[16]L.L. Hench, Sol-Gel Silica Properties, Processing and technology Transfer, *chapter 10, Biological Implications*, 116-163 (1999).
[17]P.E. Keeting, et al, *J. Bone Mineral Res.*, **7**, 1281 (1992).
[18]C.M. Botelho, R. A. Brooks, S.M. Best, M. A. Lopes, J.D. Santos, N. Rushton, W. Bonfield, Biological and Physical-Chemical Characterization of Phase Pure HA and SI-Substituted Hydroxyapatite by Different Microscopy Techniques, *Key Engineering Materials*, **254-256**, 845-848, (2004).
[19]I.R. Gibson, K.A. Hing, S.M. Best, W. Bonfield, Enhanced in vitro cell activity and surface apatite layer formation on novel silicon substituted hydroxyapatites, Ohgushi H, Hastings GW, Yoshikawa T, editors. *12th International Symposium on Ceramics in Medicine*, Nara, Japan, 191–4 (1999).

[20]C. M. Botelho, M. A. Lopes, I. R. Gibson, S. M. best, J. D. Santos, Structural analysis of Si-substituted hydroxyapatite: Zeta potential and X-Ray photoelectron spectroscopy, *Journal of Materials science: Materials in Medicine*, **13**, 1123-1127 (2002).

[21]A. Aminian, Effect of Silicon Substitution on Bioactivity of Hydrothermal Synthesized of Hydroxyapatite Nano-powders, *MSc. Thesis, Amirkabir University of Technology*, (2009).

[22]E. Landi, A. Tampieri, G. Celotti, S. Sprio, Densifcation behaviour and mechanisms of synthetic hydroxyapatites, *Journal of the European Ceramic Society*, **20**, 2377-2387 (2000).

[23]J.A. Stephen, J.M.S. Skakle, I.R. Gibson, Synthesis of Novel High Silicate-substituted Hydroxyapatite by Co-substitution Mechanisms, *Key Engineering Materials*, **330-332**, 87-90 (2007).

[24]F. Bakhshi, Biomimetic Synthesis of Nanostructured Nonstoichiometric Hydroxyapatite, *MSc. Thesis, Amirkabir University of Technology*, 2006.

[25]D. Arcos, J. Rodríguez-Carvajal, M. Vallet-Regí, The effect of the silicon incorporation on the hydroxylapatite structure A neutron diffraction study, Solid State Sciences, **6**, 987–994 (2004).

[26]J. W. Reid, L. Tuck, M. Sayer, K. Fargo, J. A. Hendry, Synthesis and characterization of single-phase silicon-substituted a-tricalcium phosphate, *Biomaterials*, **27**, 2916–2925 (2006).

[27]A.E. Porter, S.M. Best, W. Bonfield, Ultrastructural Charaterisation of Hydroxyapatite and Silicon-Substituted Hydroxyapatite, *Key Engineering Materials*, **240-24**, 505-508 (2003).

[28]S.V. Dorozhkin, E.I. Dorozhkina, F.N. Oktar, S. Salman, A Simplified Preparation Method of Silicon-Substituted Calcium Phosphates According to Green Chemistry Principles, *Key Engineering Materials*, **330-332**, 55-58 (2007).

Bioglasses
and Glass-Ceramics

ANTIBACTERIAL Ag-DOPED GLASS-CERAMIC SCAFFOLDS

Enrica Vernè[1], Marta Miola[1], Cristina Balagna[1], Giacomo Fucale[2], Giovanni Maina[3], Rosa Angela Canuto[4], Silvia Saracino[4], Giuliana Muzio[4] and Chiara Vitale-Brovarone[1]

[1] Materials Science and Chemical Engineering Dep.- Politecnico di Torino, Turin, Italy
[2] Chemical, Clinical and Microbiological Laboratory of C.T.O., Turin, Italy
[3] Traumathology and Work Medicine Department of C.T.O., Turin, Italy
[4] Department of Experimental Medicine and Oncology, University of Turin, Italy

ABSTRACT

A bioactive and fluoroapatite-containing glass-ceramic was used to realize scaffolds (Fa-GC) with the polymeric sponge method. In order to confer antibacterial properties, silver was introduced in the material (Ag-Fa-GC) through a patented ion-exchange process.

Scaffolds structure, morphology and composition were characterized through XRD, SEM and EDS analyses; the total porosity, the pore size and the degree of pore interconnection were respectively investigated by means of density measurement, image analysis and capillarity test. The mechanical strength was estimated through compressive test. Released Ag ions were evaluated using the GFAAS analysis and *in vitro* tests were performed soaking Ag- and Fa-GC in SBF solution. Antibacterial properties were verified through the dilution broth test and the measurement of inhibition zone, using a *S. Aureus* standard stock. Osteoblasts cultures were realized to investigate material biocompatibility and Ag effect.

Fa-GC with an open and interconnected porosity and a compression strength of 2.0 ± 0.6 MPa were successfully obtained. Ag-Fa-GC maintain their bioactivity after the ion-exchange treatment and the introduced Ag was gradually released in SBF. The Ag-Fa-GC showed a good antibacterial behaviour, since they are able to create a significant inhibition halo and limit the bacterial proliferation. Nevertheless the preliminary cytotoxicity test showed that Ag presence affects cells proliferation and viability.

INTRODUCTION

A good scaffold for bone tissue engineering should be characterized by an interconnected porosity greater than 50-60%, a structure composed by macropores with dimensions between 100 and 500μm and micropores of about 10μm suitable for cells colonization and proliferation, and mechanical properties similar to human bone. Moreover, a synthetic bone substitute should be preferably easily shaped in order to better fit into the bone defect.

Recently, ceramics, glasses and glass-ceramics have been largely investigated because of their ability to form *in vivo* an apatite layer able to promote the osteointegration. In particular, the interest about glasses and glass-ceramics increases because their composition could be changed in order to better tailor the reactivity in the human body. In fact, on the basis of their application, these materials can be slightly bioactive or completely bioresorbable, they are able to release ions supportive for bone mineralization and can contain a specific crystalline phases (i.e. apatite) similar to inorganic phase of human bone.

On the other hand, the infections development remains one of the most important problem after a surgical implant, in particular in orthopaedic surgery. In fact, an infection can cause prolonged hospitalization periods, patients diseases and medical costs increasing [1,2].

Silver is a well known antimicrobial agent and it is already used in medical field [3,4]; in orthopaedics, many research studies have been carried out to prevent infection on orthopaedic implants, in particular on fixation devices for osteosynthesis [5-8]. However the major risk related to silver use is the potential toxicity not only for bacteria but also for tissue cells. But it is possible to find a "safety range" of Ag concentration with a toxic effect for bacteria and a non toxic one for tissue cells, as silver effect depends on its concentration and on cells dimension (oligodynamic effect).

The aim of this work is to realise biocompatible, bioactive glass–ceramic scaffolds for bone substitution with antibacterial properties, by introducing silver ions into the scaffolds surfaces through a patented ion-exchange process [10]. This technique allows a controlled silver introduction in the superficial layers of the scaffold, maintaining unchanged its structure and its characteristics. In such a way, an effective osteointegration due to the 3D-structure of the scaffold and of its bioactivity can be obtained.

MATERIALS AND METHODS

The used glass-ceramic has the following molar composition: 50% SiO_2, 18% CaO, 9% CaF_2, 7% Na_2O, 7% K_2O, 6% P_2O_5 and 3% MgO. After melting and quenching process, glass-ceramic crystalline structure was composed by fluoroapatite crystals suitable for bone substitution, shown by means X-ray diffraction. So the material is labelled Fa-GC due to its glass-ceramic nature and the presence of fluoroapatite crystals.

The scaffolds were prepared with the polymeric sponge method where a polyurethane sponge was impregnated with a slurry containing Fa-GC powders, polyvinyl alcohol as a binder and distilled water and then thermally treated.

The obtained scaffolds have been characterized by means scanning electron microscopy and energy dispersion spectrometry (SEM-EDS Philips 525M) to evaluate the degree of sintering, the pore interconnection and size, both on the scaffold surface and on its cross-section. Moreover, the micro and macro-porosity degree and interconnection have been evaluated by means of capillary test and image analysis (software Qwin Leica). A compressive test with an Instron machine was utilized to analyse the mechanical strength of the scaffolds; the compressive strength was calculated dividing the peak load for the resistant section of tested cubic scaffolds.

The ion-exchange technique was used in order to confer antibacterial properties to the prepared scaffolds. This process was applied by the authors to bulk materials and to glass-ceramic coatings [10-12] and allows the loading silver into the material surface, without altering its characteristics, such as bioactivity and mechanical properties. The technique is based on an exchange between modifier cations of the glass-ceramic and silver ions, coming from the exchange solution. The treatment was carried out in an aqueous solution of 0,05M silver nitrate maintained at 37°C for 8 hours [13].

Scaffolds sterilization was carried out in a furnace at 180°C for 3 hours and the sterilized silver doped scaffolds will be named Ag-Fa-GC.

To evaluate silver release, three Ag-Fa-GC were soaked in 30 ml of simulated body fluid (SBF, proposed by Kokubo [14]) and incubated at 37°C up to 28 days. 1 ml of SBF was spilled from the soaking solution respectively after 3 hour, 1, 3, 7, 14, 28 days of soaking. All the spilled solutions were analysed in a graphite furnace atomic absorption spectroscopy (GFAAS PERKIN ELMER - mod. 4100 ZL).

Antibacterial properties were verified through the measurement of inhibition zone, the dilution broth test and the evaluation of the broth turbidity through McFarland index measurements. In particular, the test in broth quantified bacteria adhesion on the scaffold surfaces and their proliferation in the culture broth, through the count of bacteria colonies forming units (CFU). These tests were performed in accordance to NCCLS standards [15,16] using *Staphylococcus Aureus* standard stock (ATCC 29213). First of all, a bacterial broth was prepared through the dissolution of a *S. Aureus* lyophilized disk in 5 ml of brain-heart infusion; after 24 h of incubation at 37°C, 10 μl of suspension were spread on blood-agar plate and incubated 24 h in order to allow the bacterial colonies to growth. Afterwards, some bacterial colonies were dissolved in a physiological solution to obtain a standard 0,5 McFarland solution; the McFarland index is the optical measure of a solution turbidity (Phoenix 237 Spec BD McFarland), which supplies an indication of bacteria proliferation, since the turbidity of a solution increases with the bacteria proliferation.

Three Ag-Fa-GC and one scaffold as control were leaned on Mueller Hinton agar plate after spreading an aliquot of 0.5 McFarland suspension, and they were incubated at 37°C. After 24h of incubation, the inhibition halo was observed and measured. Moreover three Ag-Fa-GC and three control scaffolds were individually introduced in a test-tube containing a solution with a bacteria concentration of about $5 \cdot 10^5$ CFU/ml, composed by 5 ml of Mueller Hinton broth and 25 μl of 0,5 McFarland solution. All test-tubes were incubated at 37°C and after 24h, McFarland index of the broth was measured again to verify the turbidity degree. In case of bacteria proliferation, the broth turbidity will increase leading to a higher McFarland index. After the test, the samples were rinsed in 3ml of physiological solution (0.9% P/V of NaCl in distilled water) and vortexed in 3ml of physiological solution for 1 minute at 50 Hz. The bacterial broth, the washing solution and the vortexing one were analysed in order to count the CFU proliferated in the broth and adhered on scaffolds surface. Serial dilutions of all solutions were carried out to better count the CFU on the plates and an aliquot of the last one was spread on a Blood-agar plate. The number of dilution was chosen on the basis of previous works [17,18]. CFU were counted on the Blood-agar plate after 24h of incubation at 37°C.

The biocompatibility of Fa-GC and Ag-Fa-GC was evaluated using human osteoblast-like cells (MG-63), in order to estimate their behaviour on a 3D-structure and to assess the effects of the introduced silver. A 3D-structure is characterised by a greater surface reactivity due to the presence of a high number of micro and macropores. Control scaffolds, previously dry sterilized at 180°C for 3 hours, and Ag-Fa-CG have been placed in multi-wells and soaked in 12 ml of MEM culture medium (M5650 Sigma, Minimum Essential Medium) added with 1mM sodium pyruvate, 1% antibiotics and antimycotics solutions, 2 mM glutamine and 10% serum. After 24 hours, 90000 cells were seeded on the scaffolds and maintained for 2, 4, 8, 10 and 20 days at 37°C in atmosphere containing 5% of CO_2 and 95% of air. The medium was changed each 4 days. At the end of each incubation period, several tests have been carried out to estimate the material biocompatibility: MG-63 count, viability tests, such as Trypan blue exclusion test and lactate dehydrogenase (LDH) measurement, pH measurement, SEM observation and calcium nodules evaluation. All tests were performed in triplicate.

RESULTS

The polymeric sponge as template was impregnated with slurry of optimized characteristics containing 6 wt.% of PVA, 35 wt.% of Fa-GC powder and the rest distilled water; the impregnation parameters and glass-ceramic particles size were tailored in order to obtain a better covering of the polymeric skeleton and so a good sintering.

The impregnated sponge was thermally treated at 700 °C for 1 h, to allow the burnout of the polymeric template and the complete sintering of Fa-GC particles obtaining 3D-scaffolds; the sintering temperature was chosen on the basis of hot stage microscopy results (not reported). Figure 1 reports a Fa-GC scaffold image and a SEM micrograph; as it can bee observed the Fa-GC structure is very similar to the cancellous bone one, with a continuous 3D-interpenetrating network of struts and open pores between 100 and 500 μm, suitable for cell penetration and migration, and micropores (10-40 μm) useful for the vascularization and nutrient supply.

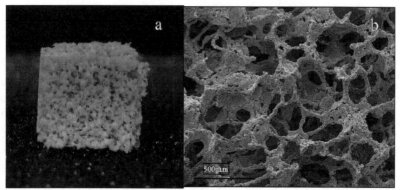

Figure 1: Fa-GC scaffold image (a) and its SEM micrograph (b).

XRD analyses (not reported) lead to the identification of fluoroapatite and of a second phase identified as canasite [(Na,K)$_6$Ca$_5$Si$_{12}$O$_{30}$(OH,F)$_4$]. The presence of canasite in the glass–ceramic is a positive attribute as this phase is highly biocompatible and has been proposed for glass–ceramic for hard-tissue augmentation.

After sintering process, the scaffolds showed a volumetric shrinkages of 62 ± 2%, this parameter can be controlled in order to realize scaffolds of precise size and profile by carefully shaping the polymeric template. Mechanical test reports a compressive strength of 2 ± 0,6 MPa, that represents a good balance together with an average porosity of 74.6 ± 3,4 % for a bone grafts if compared to data reported in literature and in the ISO Standard (ISO 13779-1:2000).

The capillarity test carried out on a 1 cm^3 cubic scaffold showed an up-take time of 2 s during which the whole scaffold became red due to the presence of ink drops in the simulated blood fluid.

Silver ions were introduced by means of a patented ion-exchange process [10], in order to confer antibacterial properties to the scaffolds,. EDS analyses performed on silver doped samples showed the presence of Ag both in the scaffold surface (1,7 at.%) and on its cross- section (0,6 at.%) after the ion-exchange.

XRD analysis and SEM observations on Ag-Fa-GC (not reported) demonstrated that the silver presence did not induce any new crystallization phenomena or morphological modifications. Silver release trend is reported in figure 2. As it can be observed, Ag$^+$ ions were gradually released in SBF and any plateau was found for the tested time frames (up to 28 days). This trend could indicate that a considerable silver amount remained on the scaffolds pore wall surface and would be released at longer times. Future tests will investigate this feature and will evaluate the release kinetics for longer times.

Nevertheless, the higher amount of silver was released during the first hours and days of immersion; this is an interesting feature, since the first time frames are the most critical for infections development, while a subsequent slowly Ag release can be important in order to maintain an antibacterial effect in case of latent infections. However, the total released silver amount should be under the toxic level, since it can be accumulated on neo-formed tissue, causing toxic effects toward cells.

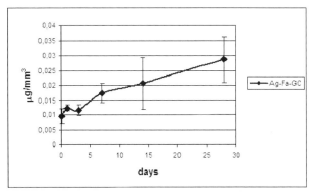

Figure 2: silver release trend in SBF solution.

The antibacterial behaviour of the Ag-Fa-GC was assessed by the inhibition halo tests, the CFU counts and by the McFarland index evaluation. The zone of inhibition test demonstrated the antimicrobial properties of Ag doped samples: silver-containing scaffolds produced a significant halo of about 3 mm, showing the effectiveness of the ion-exchange process to induce a bacteriostatic behaviour towards *S. Aureus* stock (not reported). The McFarland index evaluation showed a clear reduction of solution turbidity and so of bacterial proliferation (Figure 3).

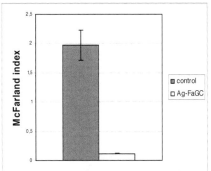

Figure 3: McFarland index values for control scaffold and Ag-Fa-GC.

A careful quantification of the bacteria growth and adhesion on the scaffold surfaces was carried out with the CFU count. Table 1 reports the CFU proliferated in the broth surrounding the scaffolds, in the liquid film in direct contact with the samples and the ones adherent onto the scaffolds. The silver presence was able to decrease the bacteria number of 2-3 magnitude order respect to the control scaffolds and so about 99.9% (3-Log reduction) of bacterial colonies did not proliferate.

Table 1: CFU adhered and proliferated.

	Control	Ag-Fa-GC
Broth	10,4±0,2	7,4±0,1
Wash	8,5±0,1	6,3±0,1
Vortex	8,0±0,1	4,7±0,1

Figure 4 shows the comparison of blood-agar plates of the control scaffold on the left and of the Ag containing Fa-GC scaffolds on the right. The effect of silver in reducing the bacterial colonies adhesion and proliferation is really powerful; on the basis of the obtained results, it is difficult to distinguish between Ag bactericidal and bacteriostatic effect.

Nevertheless, the aim of this work was to strongly reduce the proliferation of *S. Aureus* and the related risk of infection onset and thus, the observed bactericidal effect satisfied the proposed needs.

Figure 4: CFU on blood-agar plates of the control scaffold (left) and of the Ag- Fa-GC (right).

The biocompatibility of Fa-GC scaffolds and silver effect on cell proliferation were evaluated by counting the number of cells inside and outside the control scaffold and Ag-Fa-GC and those floating in the medium.

Control scaffolds were able to allow the cell proliferation inside and outside the scaffold reaching the highest peak of proliferation at 10 and 8 experimental days, respectively. Then, cell proliferation decreased for both inside and outside cells. But while inside cells were all alive, as evidenced by trypan blue exclusion test, outside cells were dead for 4% at 10 days and 20% at 20 days (data not shown). It is possible that the cells inside the scaffold blocked the proliferation, because of the increased activity in bone deposition. The presence of dead cells outside the scaffold was confirmed by the increased number of cells detached from the monolayer and floating in the medium.

Instead, Ag concentration used in this experimental protocol affected cell proliferation inside and outside the scaffold, detaching the cells from the monolayer already in the first experimental times. This detrimental behaviour was likely due to the accumulation of released silver in the cells suspension until 4 days of treatment, after that the medium was renewed, but the cells were not able to recover cell proliferation.

pH evaluation test did not evidence any differences in the control and silver doped one. Lactate dehydrogenase release showed an increase during the experimental time only in the controls, whereas it was not found in the presence of Ag. This feature seemed unusual as cells die, but in the literature it was described that Ag could be an inhibitor of lactate dehydrogenase [19] and probably the reason of lack of enzymatic activity. In conclusion, both cell viability tests specified that cells appeared dead for necrosis.

Alizarin red S staining showed great differences between the as done Fa-GC scaffolds and the ones doped with Ag. In fact cells cultures on control scaffolds were very active as they induced the deposits of calcium, whereas the calcium deposits were very few or absent with the presence of Ag.

SEM micrographs of osteoblasts cultured demonstrate the low number of cells attached on Ag-doped scaffolds: it was very difficult to find attached cells on scaffolds walls. On the other hand,

SEM micrograph of cells cultured on control scaffolds show a lot of well spread MG-63 with many filopodes bridges on the surfaces at all the studied culturing times, as shown in figure 5.

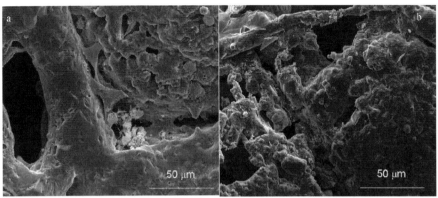

Figure 5: SEM micrographs of osteoblasts cultured for 20 days on control scaffold(a) and on Ag-Fa-GC (b).

In conclusion, it is possible to state that the Ag concentrations used in this research can be toxic not only for the bacteria, but also for the cells; nevertheless, the data suggest that the amount of released silver is quite low and should not affect cells viability.

The observed cytotoxic effect could be related to the settings of *in vitro* tests that did not simulate the fluids renewal; for this reason, future experiments should concentrate on a better setting of cells culture experiments that would take into account the renewal of physiological fluids also in the first experimental times.

CONCLUSION

The sponge impregnation method used in this research work represents a valid processing technique as it allows producing scaffolds with properties similar to cancellous bone.

The ion-exchange process allows a controlled silver introduction and a prolonged release, up to 1 month; silver ions are released in greater amount during the first hours and days which are considered the most critical for infections after a surgical treatment.

Biological tests confirms the good antibacterial effect of the Ag-containing scaffolds towards *Staphylococcus Aureus* stock, both in terms of bacteria adhesion and proliferation.

Nevertheless, cells culture tests shows that Ag presence negatively affected osteoblast-like cells proliferation, limiting cells growth and viability although the total released silver is well under the toxic limits. Future works will focus on scaffolds doped with lower Ag amounts, by tailoring the ion-exchange parameters, and on an optimization of the cells culture experiments, in order to reach a concentration toxic for bacterial but not for tissue cells.

REFERENCES

[1] G.W Green, R.P. Wenzel, Postoperative wound infections: a controlled study of the increased duration of hospital stay and direct cost of hospitalization, *Ann. Surg.*, **185**, 264 (1977)
[2] J.D. Whitehouse, N.D. Friedman, K.B. Kirkland, W.J. Richardson, D.J. Sexton, The impact of surgical-site infections following orthopaedic surgery at a community hospital and a university hospital, adverse quality of life, excess length of stay, and extra cost., *Infect Control Hosp.*, Epidemiol. **23**, 183 (2002)

[3] B.S. Atiyeh, M. Costagliola, S.N. Hayek, S.A. Dibo, Effect of silver on burn wound infection control and healing: review of literature, *Burns, 33*, 139 (2007)

[4] J.B. Wright, D.L. Hansen, R.E. Burrel, The comparative efficacy of two antimicrobial barrier dressings: in vitro examination of two controlled release of silver dressings, *Wounds, 10(6)*, 179 (1998)

[5] M.A. Wassal, M. Santin, C. Isalberti, M. Cannas, M.P. Denyer, J. Biomed. Adhesion of bacteria to stainless steel and Silver-coated orthopedic external fixation pins, *Mater. Res.*, 36, 325 (1997)

[6] A. Massè, A. Bruno, M. Boretti, A. Biasetti, M. Cannas, P. Gallinaro, J. Biomed. Prevention of pin track infection in external fixation with Silver coated pins: clinical and microbiological results, *Mater. Res.*, 53 (5), 600 (2000)

[7] G. Gosheger, J. Hardes, H. Ahrens, A. Streitburger, H. Buerger, M. Erren, A. Gunsel, F.H. Kemper, W. Winkelmann, C. Eiff, Silver-coated megaendoprostheses in a rabbit model: an analysis of the infection rate and toxicological side effects, *Biomat.* 25, 5547 (2005)

[8] M. Bosetti, A. Massè, E. Tobin, M. Cannas, Silver coated biomaterials for external fixation devices: in vitro biocompatibility and genotoxicity. *Biomat.* 23, 887 (2002)

[9] S.F. Hulbert, S.J. Morrison, J.J. Klawitter, Tissue reaction to three ceramics of porous and non-porous structures J. Biomed. Mater. Res. 6, 347 (1972)

[10] S. Di Nunzio, E. Verne', Process for the production of silver-containing prosthetic devices. Patent WO 2006/058906

[11] S. Di Nunzio, C. Vitale Brovarone, S. Spriano, D. Milanese, E. Verne', V. Bergo, G. Maina, P. Spinelli, Silver containing bioactive glasses prepared by molten salt-ion exchange, *J. Eur. Ceram. Soc.*, 24, 2935 (2004)

[12] E. Verne', S. Di Nunzio, M. Borsetti, P. Appendino, C. Vitale Brovarone, G. Maina, M. Cannas, Surface characterization of silver-doped bioactive glass, *Biomater.*, 26/25, 5111 (2005)

[13] C. Vitale Brovarone, M. Miola, C. Balagna, E. Vernè, 3D-glass–ceramic antibacterial scaffolds with properties for bone grafting, *Chem. Eng. J.* 137, 129 (2008)

[14] T. Kokubo, H. Takadama, How useful is SBF impredicting in vivo bone bioactivity?, *Biomaterials, 27*, 2907 (2006)

[15] National Committee for Clinical Laboratory Standards (NCCLS) M2-A9 "Performance Standards for Antimicrobial Disk Susce.ptibility Tests, Approved Standard – Ninth Edition".

[16] National Committee for Clinical Laboratory Standards (NCCLS) M7-A6 "Methods for Dilution Antimicrobial Susceptibility Tests for Bacteria that grow Aerobically, Approved standard 6th ed.

[17] E. Verné, S. Ferraris, M. Miola, G. Fucale, G. Maina, G. Martinasso, R.A. Canuto, S. Di Nunzio, C. Vitale Brovarone, Synthesis and characterization of a bioactive and antibacterial glass-ceramic (II): Plasma Spray coatings on metallic substrates, *Adv. Appl. Cer.*, 107 (5), 234 (2008)

[18] E. Verné, S. Ferraris, M. Miola, G. Fucale, G. Maina, P. Robotti, C. Bianchi, G. Martinasso, R.A. Canuto, C. Vitale Brovarone, Synthesis and characterization of a bioactive and antibacterial glass-ceramic (I): microstructure, properties and biological behaviour. Advances In Applied Ceramics, *Adv. Appl. Cer.* 107 (5), 245 (2008)

[19] P.M. Manchery, C.E. Wright, A radiotracer probe to study metal interaction with human lactate dehydrogenase isoenzymes, J. Protein Chemistry 8 (6), 757 (1989)

BIOACTIVE GLASS FOR BONE AND JOINT REPAIR

Mohamed N. Rahaman[1,*], Qiang Fu[1], B. Sonny Bal[2], Delbert E. Day[1], and Hailuo Fu[1]
[1]Missouri University of Science and Technology, Department of Materials Science and Engineering, and Center for Bone and Tissue Repair and Regeneration, Rolla, Missouri 65409, USA
[2]Department of Orthopaedic Surgery, University of Missouri-Columbia, Columbia, Missouri 65211, USA
*Corresponding author; e-mail: rahaman@mst.edu

ABSTRACT
This paper provides a review of some of our recent work involving the use of bioactive glasses in engineering hard and soft tissues for bone and joint repair in humans and animals. Bioactive glasses have held great promise for bone repair and regeneration, but their limitations have prevented them from realizing their potential. Recent research, utilizing new or modified bioactive glass compositions and novel processing techniques, has shown potential for overcoming these limitations, while maintaining the appealing properties of bioactive glasses. Our recent work has shown the ability to form porous, three-dimensional bioactive glass scaffolds with novel microstructures and unique mechanical properties for potential use as a segmental bone defect replacement or bone grafting material. Special compositions of bioactive glasses have been converted into novel biocompatible structures, such as hydroxyapatite microspheres and fibers, hollow or solid, which have potential application in drug or protein growth factor delivery for the repair and regeneration of hard tissues. Bioactive glasses have also shown excellent potential recently for use as a media supplement to enhance *in vitro* culture of tissue-engineered cartilage, and as a substrate in tissue-engineered osteochondral constructs for the healing of experimental cartilage defects in adult rabbit knees. The potential benefits of bioactive glasses in engineering hard and soft tissues for bone and joint repair are discussed.

INTRODUCTION
Since the discovery by Hench et al.[1] in 1971 that certain silicate glasses reacted with the body fluids in humans and animals to form a strong bond with hard and soft tissues, 'bioactive' glasses have held great promise for applications in bone repair. Subsequent research showed that bioactive glasses reacted to form a hydroxyapatite-like phase, similar in composition to the main mineral constituent in living bone, which is responsible for the strong chemical bond between the silicate bioactive glass and bone.[2] Bioactive glasses have a widely recognized ability to support the growth of bone cells,[3,4] and they are reported to release ions that activate expression of osteogenic genes.[5,6] More recent work indicate that bioactive glasses have the ability to stimulate angiogenesis.[7,8] The bioactivity of glass, as measured by the rate at which it reacts to form the hydroxyapatite (HA) phase, can be varied over a wide range, from hours to months, depending on the glass composition.[9,10] Because of the compositional and fabrication flexibility of glass, scaffolds with a wide range of chemical and physical properties as well as porous architectures can be prepared.

Despite the appealing properties of bioactive glasses, certain limitations have prevented these materials from realizing their potential in bone tissue engineering applications. The demand for porous synthetic scaffolds to repair damaged or diseased bone is increasing because of concerns associated with current treatments using bone autograft or allograft. However, bioactive glasses are brittle, and porous three-dimensional scaffolds of bioactive glass often lack the requisite combination of high strength and high porosity for skeletal substitution of load-bearing bones. They are limited to low-stress applications instead, such as filling of contained bone defects, where adjacent, intact bone provides mechanical rigidity and support. Some bioactive glasses, such as the well-recognized silicate 45S5 composition discovered by Hench et al.,[1] convert slowly and incompletely to HA. Silicate 45S5 glass is also difficult to form into porous scaffolds with anatomically relevant shapes because of limited thermal bonding of the glass due to insufficient viscous flow or to crystallization of the glass.

Recent work by the authors, utilizing new or modified bioactive glass compositions and novel forming methods have shown potential for overcoming some of the limitations of bioactive for bone repair applications, while maintain their appealing properties. In particular, our work has shown the ability to form bioactive glass scaffolds with controllable bioactivity,[11–13] anatomically relevant shapes,[14] novel pore architectures,[15,16] high strength, and high strain for failure.[16] Bioactive glasses have also been converted into novel structures, such as HA microspheres and fibers, hollow or solid, [17–21] which have potential applications as devices for drug or growth factor delivery in tissue engineering. Bioactive glass has also been shown to have potential as a media supplement to enhance *in vitro* culture of tissue-engineered cartilage[22] and as a substrate in tissue-engineered osteochondral constructs for the healing of cartilage defects in rabbit knees.[23] The objective of this paper is to review our recent work on the use of bioactive glasses in engineering hard and soft tissues for bone and joint repair.

BIOACTIVE GLASS SCAFFOLDS FOR BONE REPAIR

A silicate bioactive glass,[24] designated 13-93, with the composition given in **Table 1**, was prepared by melting a mixture of analytical grade Na_2CO_3, K_2CO_3, $MgCO_3$, $CaCO_3$, SiO_2 and $NaH_2PO_4.2H_2O$ (Fisher Scientific, St. Louis, MO) in a platinum crucible at 1300°C, casting the melt on a cold metal plate, and grinding the glass to particles <5 μm by attrition milling. This glass was used because of its well-documented bioactivity, and its ability to be thermally bonded into three-dimensional (3-D) structures with anatomically relevant shapes without crystallization of the glass. The glass is also approved for *in vivo* use in the United States and Europe.

Scaffolds with a 'trabecular' microstructure were prepared by a polymer foam replication technique, as described in detail elsewhere.[15] Briefly, a commercial polymer foam (~10 mm in diameter × 20 mm) with a pore architecture similar to that of dry human trabecular bone was infiltrated with an aqueous slurry of glass particles (35 vol%). After drying, the as-coated foam was heated to decompose the polymer phase, and sintered (1 h at 700°C) to bond the glass particles into a dense network. Scanning electron microscopy (SEM) confirmed that the scaffolds prepared by this polymer foam replication technique (**Fig. 1a**) had a microstructure similar to that of human trabecular bone (**Fig. 1b**). The microstructure consisted of a dense network of glass, and interconnected cellular pores (porosity = 85 ± 2%; pore size = 100–500 μm). The pore characteristics can be varied, while maintaining a trabecular microstructure, by manipulating the particle concentration of the slurry, the infiltration process, and the sintering conditions. Repeating the slurry infiltration step or using a higher sintering temperature (or longer sintering time) leads to a reduction in porosity and average pore size, and to higher strength. Trabecular scaffolds with porosities in the range 55–90% have been prepared using this technique.

Scaffolds with oriented pore architectures were prepared by a unidirectional freeze casting method.[25–28] Suspensions of 13-93 bioactive glass particles (10–15 vol%) containing aqueous solvents with or without 1,4–dioxane were frozen unidirectionally on a cold substrate.[16] After sublimation of the frozen liquid in a freeze dryer, the scaffolds were sintered under the same conditions described for the trabecular scaffolds. Unidirectional freezing of aqueous suspensions resulted in scaffolds with a lamellar microstructure (not shown), in which the pores were oriented in the direction of freezing. The lamellar scaffolds had pore widths in the range 10–40 μm, which were smaller than the minimum pore size (~100 μm) reported for scaffolds capable of supporting cell proliferation and function.[29,30]

Table 1. Compositions of bioactive glasses, designated 13-93 and CaLB15, used in this investigation, and the composition of the widely-studied 45S5 bioactive glass

Composition (wt%)	13-93	CaLB15	45S5
Li_2O	0	10.7	0
Na_2O	6.0	0	24.5
K_2O	12.0	0	0
MgO	5.0	0	0
CaO	20.0	15.0	24.5
SiO_2	53.0	0	45.0
P_2O_5	4.0	0	6.0
B_2O_3	0	74.3	0

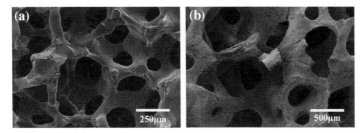

Fig. 1. SEM images of (a) 13-93 bioactive glass scaffolds fabricated by a polymer foam replication technique, and (b) dry human trabecular bone.

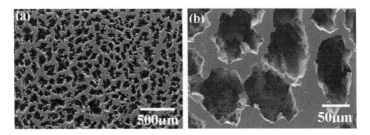

Fig. 2. SEM images of the cross section of 13-93 bioactive glass scaffold prepared by unidirectional freezing of suspensions (15 vol% particles) containing a solvent of water + 60 wt% dioxane. The cross section is perpendicular to the freezing direction.

Unidirectional freezing of aqueous suspensions containing 40–70 wt% dioxane resulted in a marked changed of the oriented microstructure, giving columnar pores oriented in the direction of freezing. The width of the columnar pores varied in a non-linear manner with dioxane concentration, with the largest pore width obtained at a dioxane concentration of ~60 wt%. **Figure 2** shows an SEM image of a bioactive glass scaffold sectioned perpendicular to the freezing direction, which was prepared from a suspension (15 vol% particles) with 60 wt% dioxane. The scaffolds had a porosity of 65 ± 2%, and pore widths of 90–110 μm, which satisfied or exceeded the minimum requirements indicated earlier for scaffolds capable of supporting cell proliferation and function.

X-ray diffraction (XRD) showed that under the sintering conditions described earlier, the as-prepared 13-93 bioactive glass scaffolds remained amorphous. Immersion of the scaffolds in a simulated body fluid (SBF) resulted in a specific surface reaction, leading to the formation of a porous layer of nanometer-sized, needle-like crystals on the glass surface in less than 7 days (**Fig. 3**). The XRD pattern of the converted surface layer corresponded to that of a reference HA (JCPDS 72-1243), confirming the formation of a HA layer on the surface of the scaffold.

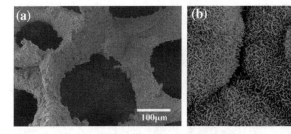

Fig. 3. SEM images of the surface of a 13-93 bioactive glass scaffolds with the trabecular microstructure after immersion for 7 days in a simulated body fluid.

The mechanical response of the trabecular and columnar scaffolds showed marked differences when tested in compression. Testing was performed according to ASTM C-773 in an Instron testing machine at a crosshead speed 0.05–5 mm/min, using specimens with a length to diameter ratio of ≥2. The stress vs. deformation behavior for the trabecular bioactive glass scaffolds (**Fig. 4a**) was linear elastic during the initial compression, followed by several peaks and valleys, presumably caused by progressive fracture of individual glass struts, and finally complete failure of the specimen. The elastic modulus (slope of the initial linear region) was 3.0 ± 0.5 GPa, while the compressive strength, taken as the highest stress on the stress vs. deformation curve, was 11 ± 1 MPa (deformation rate = 0.5 mm/min), equal to the highest strengths reported for human trabecular bone.

Figure 4b shows the stress vs. deformation response for 13-93 bioactive glass scaffolds with the columnar microstructure. Instead of a brittle response typical of ceramics and glasses, the scaffolds showed an elastic–plastic response that is more typical of natural materials. The stress vs. deformation curve consisted of three regions: an approximately linear region at low deformation, a region of peak stress, and a plateau region (or a region of slowly decreasing stress) at higher deformation. The scaffolds showed a compressive strength (deformation rate = 0.5 mm/min) of 25 ± 3 MPa, and a large strain for failure (>20%). The response of the scaffolds also showed strong strain-rate dependence. With increasing deformation rate, the peak stress increased and shifted to lower strain.

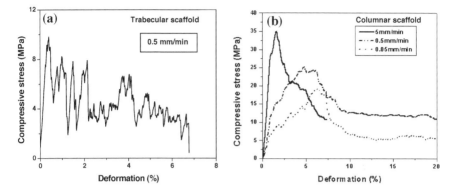

Fig. 4. Compressive stress vs. deformation for 13-93 bioactive glass scaffolds tested at the deformation rates shown: (a) trabecular microstructure; and (b) columnar microstructure (testing performed in the direction of pore orientation).

In vitro cell culture experiments showed that both the trabecular scaffolds and the columnar scaffolds effectively supported the proliferation and function of murine osteoblastic cells (MC3T3-E1 or MLO-A5). Bioactive glass (13-93) scaffolds were dry heat sterilized, seeded with 50,000 cells, placed in αMEM medium with 5% calf serum and 5% fetal calf serum, and incubated for 2 to 12 days at 37°C. The morphology of cells seeded on the scaffolds was examined by scanning electron microscopy (SEM). The SEM images in **Fig. 5** reveal complete coverage of the scaffolds by day 6. Total protein in lysates recovered from cell-seeded samples was measured with a micro-BCA reagent to assess cell growth on the scaffolds. The results (not shown) indicate a nearly linear increase in total protein during the 6 day incubation for both the trabecular and the columnar scaffolds. The protein assay results, together with the SEM images, attest to the ability of both groups of scaffolds to effectively support cell proliferation.

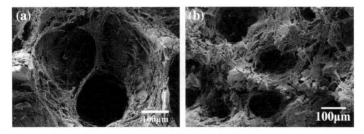

Fig. 5. SEM images of 13-93 bioactive glass scaffolds seeded with cells and cultured for 6 days: (a) trabecular scaffold seeded with MC3Te-E1 cells; (b) columnar scaffold seeded with MLO-A5 cells.

To visualize the metabolically active cells on and within the porous scaffolds, the cell-seeded scaffolds were treated with the tetrazolium salt MTT during the last 4 h of incubation, and the distribution of insoluble purple formazan, a product of the mitochondrial reduction of MTT by viable cells, was qualitatively assessed using a stereo microscope. The increase in intensity of the purple color with culture time indicated the proliferation of viable, metabolically active cells on both groups of scaffolds (**Figs. 6a** and **6c**). The purple formazan visible on the freeze fracture surface of the scaffolds cultured for 6 days (**Figs. 6b** and **6d**) indicated the presence of metabolically active cells within the interior of the scaffolds. The images in **Figs. 6b** and **6d** indicate that cell proliferation into the interior pores of the columnar scaffolds was slower than that for the trabecular scaffolds, presumably due to the smaller pore size and porosity. Additional images (not shown) of the columnar scaffolds indicated complete cell proliferation into these scaffolds after 15 days of incubation.

Alkaline phosphatase (ALP) activity in the lysates was assayed by spectrophotometric measurement of p-nitrophenol (pNP) released from p-nitrophenol phosphate.[31] Alkaline phosphatase activity, an indicator of osteogenic function, rose markedly in MLO-A5 cells cultured on the trabecular scaffolds (not shown) as well as on the columnar scaffolds (**Fig. 7**).

Formation of mineralized nodules by MLO-A5 cells cultured on scaffolds in 'mineralization medium' (medium supplemented with ascorbic acid and β-glycerolphosphate[32]) was quantitated by an alizarin red staining procedure described elsewhere.[33] The recovery of alizarin red-positive material increased with culture time for the trabecular scaffolds (not shown), as well as for the columnar scaffolds (**Fig. 8**), providing additional evidence that these scaffolds are capable of supporting differentiation of osteoprogenitor cells to functional bone tissue.

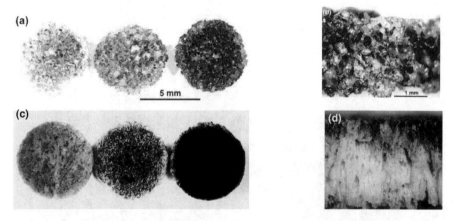

Fig. 6. Cell seeded 13-93 glass scaffolds treated with MTT: surface of (a) trabecular scaffolds, and (b) columnar scaffolds after culture intervals of 2, 4, and 6 days (left to right, respectively); (b, d) freeze fracture surface of the of the corresponding scaffolds cultured for 6 days, showing MTT-labeled cells within the interior.

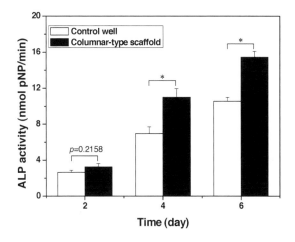

Fig. 7. Alkaline phosphatase activity in MLO-A5 cells cultured on columnar scaffolds and in control wells for 2, 4 and 6 days. Mean ± sd; n = 4. *Significant increase with increasing culture duration ($p <$ 0.05).

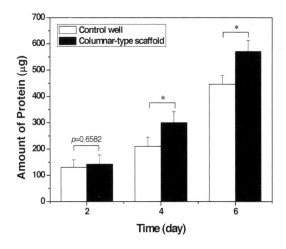

Fig. 8. Quantitative analysis of mineralization in columnar scaffolds and in control wells, by measuring the absorbance (optical density) of the extracted alizarin red staining dye using an Optima plate reader at 520 nm. Mean ± sd; n = 4. *Significant increase with increasing culture duration ($p < 0.05$).

The results show that 13-93 bioactive glass scaffolds with the trabecular microstructure or the columnar microstructure effectively support the growth and function of osteoblastic cells. These findings suggest that these porous bioactive glass scaffolds could potentially be used for bone repair and tissue engineering applications. Because of their lower strength, the trabecular scaffolds may be better suited for the repair of contained defects in which surrounding bone provides major mechanical support. On the other hand, the columnar scaffolds, with a much higher compressive strength, have potential for the repair of segmental defects in long bones.

BIOACTIVE GLASS-BASED DEVICES FOR DRUG OR GROWTH FACTOR RELEASE

Bioactive glasses with special compositions were converted into nanostructured phosphate devices using a novel process invented by Day et al.[17] This process is shown schematically in **Fig. 9**, using as an example the conversion of a Li_2O–CaO–B_2O_3 glass to form HA. After melting and forming the glass into the required shape (such as a microsphere or fiber) using conventional glass processing methods, the solid glass object is reacted in an aqueous phosphate solution at near room temperature. In the conversion process, Ca^{2+} and other ions are released as the glass dissolves, while $(PO_4)^{3-}$ and OH^- from the solution react with Ca^{2+} ions to form a porous amorphous calcium phosphate (ACP) layer on the glass surface. The glass core continues to dissolve as the ACP converts to HA, until finally a fully reacted shape, hollow or solid, composed of porous, nanostructured HA is produced. The reaction is pseudomorphic, retaining the same external shape and dimensions of the starting glass object.

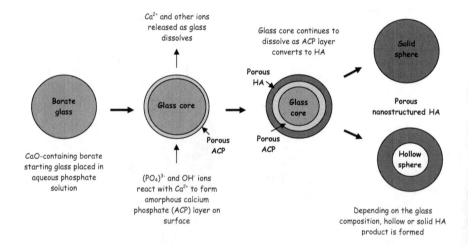

Fig. 9. Schematic diagram illustrating the conversion of a borate glass to a porous nanostructured hydroxyapatite (HA) device with a required geometry (hollow or solid microsphere).

Because of the ease of forming a glass into the required shape, phosphate devices with simple as well as complex shapes can be produced. The ability to control the shape of the phosphate device is guaranteed because of the pseudomorphic nature of the conversion process. The glass composition is a key parameter in the conversion process. It determines the rate and extent of the glass conversion, as well as the composition and structure of the phosphate product. Because some borate glasses have a lower chemical durability than the more common silicate glasses, they can be converted rapidly and completely in an aqueous phosphate solution, so they are often the preferred starting materials for the formation of phosphate devices. The concentration of CaO in the original borate glass composition determines whether a hollow or solid shape (microsphere or fiber) is formed.

Figure 10a shows microspheres (diameter = 106–150 μm) prepared from a Li_2O-CaO-B_2O_3 glass (designated CaLB3-15) with the composition given in **Table 1**. A hollow HA microsphere formed by converting the glass microspheres in a 0.02 M K_2HPO_4 solution (pH = 7.0) at 37°C is shown in **Fig. 10b**. As prepared, the hollow HA microspheres have a porous shell wall consisting of nanostructured, needle-like particles (**Fig. 10c**), with a surface area as high as 150 m^2/g.

Experiments showed promising results for using these hollow HA microspheres as devices for controlled drug or protein delivery.[19] The HA microspheres were filled with a saline solution (0.85 wt% NaCl) containing 3.5 mg bovine serum albumin (BSA) per ml by placing the microspheres in the solution and pulling a vacuum on the system. Filling of the microspheres with the BSA solution was taken to be completed when no more air bubbles were released from the microspheres. After removal from the BSA solution, the microspheres were washed with saline solution (containing no BSA) to remove surface-adsorbed BSA. The release of the BSA from the filled microspheres into a saline solution at 37°C was measured using a colorimetric protein assay.

Figure 11 shows that the release of a BSA from the hollow HA microspheres can be controlled by varying the porosity of the hollow microsphere wall. The as-formed microspheres with the highest porosity gave the fastest release of BSA, whereas the release from the microspheres sintered to 600°C was lower. The release from the microspheres heated to 900°C was much smaller, as expected, due to the vast reduction in the porosity of the microsphere wall that occurred for this higher temperature.

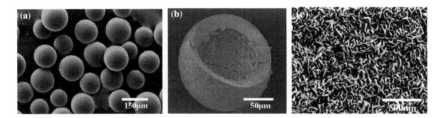

Fig. 10. SEM images of (a) Li_2O–CaO–B_2O_3 glass microspheres with a diameter of 106–150 μm; (b) hollow HA microsphere broken to reveal the void and shell wall; and (c) surface of HA microsphere consisting of needle-like nanoparticles.

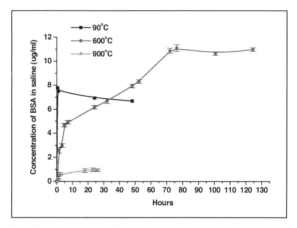

Fig. 11. Release of bovine serum albumin (BSA) from hollow HA microspheres in saline solution. The microspheres were made from borate glass and heat-treated at 90, 600, and 900°C.

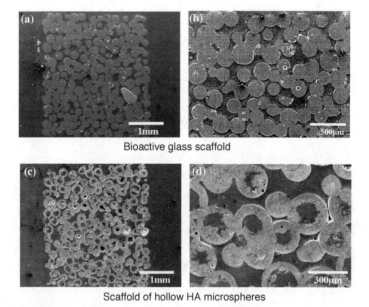

Bioactive glass scaffold

Scaffold of hollow HA microspheres

Fig 12. SEM images of cross sections of (a, b) bioactive glass scaffold formed by thermal bonding of Li_2O–CaO–B_2O_3 glass microspheres, and (c, d) scaffold consisting of hollow HA microspheres formed by conversion of the glass scaffold in an aqueous phosphate solution.

Porous three-dimensional phosphate devices were produced by thermally bonding Li_2O–CaO–B_2O_3 glass microspheres to form the desired shape, followed by conversion in an aqueous phosphate solution. As an example, CaLB15 glass particles of diameter 150–250 μm were bonded into a porous cylindrical scaffold (porosity = 45–50%) by heating for 4 h at 600°C (**Figs. 12a** and **12b**). After soaking in 0.02 M K_2HPO_4 solution for 4 days to convert the glass to HA, the scaffolds were dried, and sintered for 2 h at 560°C. **Figures 12c** and **12d** show the cross section of a cylindrical scaffold composed of hollow HA microspheres. Mechanical testing showed that the compressive strength of the scaffold was only 1.0–1.5 MPa. The strength of the scaffold is presumably limited by drying cracks after the conversion step. Current work is examining methods to improve the strength of these scaffolds.

To summarize at this stage, the hollow HA microspheres prepared by the conversion of bioactive glass in an aqueous phosphate solution show promising potential for use as devices for sustained protein growth factor release. Our recent work has shown the ability to achieve sustained release of BSA over a period exceeding 7 days. Scaffolds of hollow HA microspheres, prepared from porous 3-D constructs of thermally-bonded glass particles, could potentially be used to serve both a mechanical function as well as a protein growth factor release function.

BIOACTIVE GLASS AS A MEDIA SUPPLEMENT AND SUBSTRATE FOR TISSUE-ENGINEERED CARTILAGE

Our recent work has shown that when co-cultured with bovine chondrocyte-seeded agarose hydrogels, porous scaffolds of 13-93 bioactive glass served as a medium supplement for culturing tissue-engineered cartilage.[22] Culture was performed in serum-free, chemically defined media supplemented with transforming growth factor, TGF-β3 (10 ng/mL for the first 14 days of culture). Two studies were performed to evaluate short-term effects (2 weeks) as well as long-term effects (6 weeks) of exposing the cell-seeded hydrogels to the bioactive glass (**Fig. 13a**).

In Study 1, 13-93 bioactive glass in the form of three-dimensional scaffolds with the trabecular microstructure (**Fig. 1a**) was introduced to the culture media of agarose hydrogels on day 28 of culture, and mechanical and biochemical changes at 7 and 14 days after exposure were evaluated. In Study 2, using cells from a separate isolation, bioactive glass scaffolds were added to the culture medium immediately after cell encapsulation (and coincident with exposure of TGF-β3; days 0–14). The bioactive glass scaffolds were then removed, media was replaced, and the agarose hydrogels were cultured for 4 more weeks. Agarose hydrogels maintained in normal culture medium without bioactive glass, as well as those cultured continuously with bioactive glass for the entire culture period (6 weeks) served as controls.

Transient exposure of agarose hydrogels to 13-93 bioactive glass scaffolds resulted in higher mechanical and biochemical properties than control samples (**Fig. 13b**). In Study 1, an immediate rise in properties was seen after 7 days of exposure, with a further increase by day 14. In Study 2, significant increases in properties for cell-seeded agarose hydrogels exposed transiently to bioactive glass construct occurred by day 42. The peak values for the tissue engineered cartilage achieved in this study were an elastic modulus E_Y = 743 kPa, dynamic modulus (at 0.1 Hz) G^* = 2.8 MPa; glycosaminoglycan (GAG) content = 7.6%; and collagen contents = 6.2%. These values compare favorably with native bovine wrist cartilage. Histology of transverse slides showed more rapid and homogeneous extracellular matrix distribution for constructs exposed to bioactive glass on day 28 (**Fig. 14**). Von Kossa stains for mineralization showed no difference from control slides.

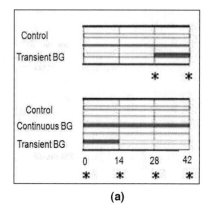

(a)

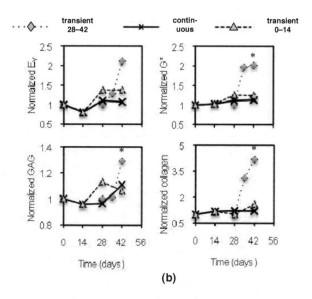

(b)

Fig. 13. (a) Schematic of the exposure of bovine chondrocyte seeded agarose hydrogels to 13-93 bioactive glass (✻ = testing days). (b) Young's modulus (E_Y), dynamic Modulus at 0.1Hz (G^*), Glycosaminoglycan (GAG), and collagen, normalized to respective controls. (*$p<0.05$ for transient response vs. continuous.) (From Ref. 22.)

GAG deposition   Collagen deposition

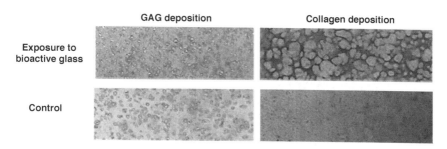

Fig. 14. Histology of transverse slides showing higher glycosaminoglycan (GAG) and collagen deposition in chondrocyte-seeded agarose hydrogels exposed to 13-93 bioactive glass on day 28. (From Ref. 22.)

Our recent work has also shown that silicate bioactive glass (13-93) can serve as a suitable substrate for tissue-engineered osteochondral constructs.[23] In this work, cylindrical plugs (3 mm in diameter × 6 mm) were made by bonding poly(ethylene glycol), PEG, hydrogel to three materials: 13-93 glass, rabbit allograft bone, and porous tantalum metal. Chondrogenic-differentiated mesenchymal stem cells (MSCs) obtained from adult allogeneic bone marrow were uniformly suspended in the 1 mm thick hydrogel cap, before implantation into experimental defects in rabbit knees (**Fig. 15**).

Examination at 6 and 12 weeks after implantation showed that both 13-93 glass and tantalum bonded to living bone, and supported viable cartilage cells at the articular layer. Osseointegration of the bioactive glass and tantalum was superior to allograft, and transplanted cells expressed type 2 collagen at 12 weeks following implantation (**Fig. 16**).

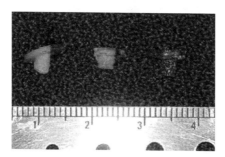

Fig. 15. Composite graft plugs (3 mm in diameter × 6 mm) made by bonding PEG hydrogel loaded with stem cell-derived cartilage cells to rabbit allograft (left), 13-93 bioactive glass (center), and porous tantalum (right). Excess hydrogel was trimmed off before implantation into matching defects created in the femoral condyles of adult rabbit knees.

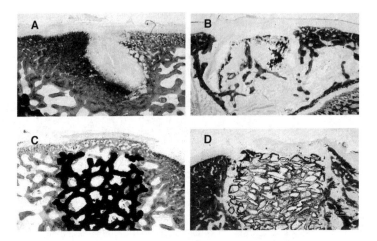

Fig. 16. Histology at 12 weeks after implantation with (A) rabbit allograft, (B) bioactive glass 13-93, (C) trabecular tantalum, and (D) control. The best osteointegration scores were obtained with bioactive glass and tantalum. Collagen type 1 and 2 immunohistochemistry also showed that surface tissue most closely resembled hyaline cartilage with these two substrate materials (data not shown). (From Ref. 23.)

SUMMARY AND CONCLUSIONS

Some recent work by the authors on the use of bioactive glass in engineering of hard and soft tissues for bone and joint repair was reviewed. Porous three-dimensional scaffolds of 13-93 glass, with a microstructure similar to that of human trabecular bone, were prepared by a polymer foam replication technique. These 'trabecular' scaffolds, with strengths >10 MPa and the ability to support proliferation and function of osteoblastic cells, could potentially be used for repairing contained defects in bones. Scaffolds of 13-93 glass with an oriented pore architecture, prepared by a unidirectional freezing of suspensions, have compressive strengths >25 MPa, and show an elastic-plastic mechanical response, with a large deformation for failure (>20%) and stain rate sensitivity. These oriented scaffolds have potential application in the repair of segmental defects in long bones. Individual microspheres of $CaO–Li_2O–B_2O_3$ glass (106–150 μm) convert rapidly (<4 days) to form hollow hydroxyapatite (HA) microspheres. These hollow HA microspheres provide sustained release of a model protein, bovine serum albumin (BSA), and could potentially be used as devices for drug or growth factor delivery in bone tissue engineering. Scaffolds composed of an interconnected network of hollow HA microspheres were formed by thermally bonding $CaO–Li_2O–B_2O_3$ glass microspheres into porous three-dimensional structures and converting the system in an aqueous phosphate solution. These unique scaffolds could provide a drug or growth factor release function in addition to a mechanical, load-bearing function. Bioactive glass also shows promise as a media supplement for *in vitro* culture of tissue engineered cartilage and as a substrate in engineering osteochondral tissues for joint repair.

ACKNOWLEDGEMENTS
This work was supported by the US Army Medical Research and Materiel Command (grant number W82XWH-08-1-0765), the National Institute of Dental and Craniofacial Research, the National Institutes of Health (grant number 1R15DE018251-01), and the Center for Bone and Tissue Repair and Regeneration (CBTRR), Missouri University of Science and Technology. The authors would like to thank Prof. C. T. Hung, Columbia University, and Prof. James L. Cook, University of Missouri-Columbia, for the results contained in Figs. 13, 14, and 16.

REFERENCES
[1]L. L. Hench, R. J. Splinter, W. C. Allen, and T. K. Greenlee, Jr., "Bonding Mechanisms at the Interface of Ceramic Prosthetic Materials," *J. Biomed. Mater. Res.*, **2**, 117–41 (1971).

[2]L. L. Hench, "Bioceramics," *J. Am. Ceram. Soc.*, **81**, 1705–28 (1998).

[3]D. L. Wheeler, K. E. Stokes, H. E. Park, and J. O. Hollinger, "Evaluation of Particulate Bioglass in a Rabbit Radius Ostectomy Model," *J. Biomed. Mater. Res.*, **35**, 249–54 (1997).

[4]D. L. Wheeler, K. E. Stokes, R. G. Hoellrich, D. L. Chamberland, and S. W. McLoughlin, "Effect of Bioactive Glass Particle Size on Osseous Regeneration of Cancellous Defects," *J. Biomed. Mater. Res.*, **41**, 527–33 (1998).

[5]I. D. Xynos, A. J. Edgar, L. D. Buttery, L. L. Hench, and J. M. Polak, "Ionic Products of Bioactive Glass Dissolution Increase Proliferation of Human Osteoblasts and Induce Insulin-Like Growth Factor II mRNA Expression and Protein Synthesis," *Biochem. Biophys. Res. Commun.*, **276**, 461–65 (2000).

[6]I. D. Xynos, A. J. Edgar, L. D. Buttery, L. L. Hench, and J. M. Polak, "Gene-Expression Profiling of Human Osteoblasts Following Treatment with Ionic Products of Bioglass 45S5 Dissolution," *J. Biomed. Mater. Res.*, **55**, 151–57 (2001).

[7]J. K. Leach, D. Kaigler, Z. Wang, P. H. Krebsbach, and D. J. Mooney, "Coating of VEGF-Releasing Scaffolds with Bioactive Glass for Angiogenesis and Bone Regeneration," *Biomaterials*, 27, 3249–55 (2006).

[8]A. Leu and J. K. Leach, "Proangiogenic Potential of a Collagen/Bioactive Glass Substrate," *Pharmaceut. Res.*, **25**, 1222–29 (2008).

[9]W. Huang, D. E. Day, K. Kittiratanapiboon, and M. N. Rahaman, "Kinetics and Mechanisms of the Conversion of Silicate (45S5), Borate, and Borosilicate Glasses to Hydroxyapatite in Dilute Phosphate Solutions," *J. Mater. Sci.: Mater. Med.*, **17**, 583–96 (2006).

[10]A. Yao, D. Wang, W. Huang, Q. Fu, M. N. Rahaman, and D. E. Day, "In Vitro Bioactive Characteristics of Borate-Based Glasses with Controllable Degradation Behavior," *J. Am. Ceram. Soc.*, **90**, 303–6 (2007).

[11]W. Liang, C. Rüssel, D. E. Day, and G. Völksch, "Bioactive Comparison of a Borate, Phosphate, and Silicate Glass," *J. Mater. Res.*, **21**, 125–31 (2006).

[12]W. Liang, M. N. Rahaman, D. E. Day, N. W. Marion, G. C. Reilly, and J. J. Mao, "Bioactive Borate Glass Scaffold for Bone Tissue Engineering," *J. Non-Cryst. Solids*, 354, 1690–96 (2008).

[13]M. N. Rahaman, R. F. Brown, B. S. Bal, and D. E. Day, "Bioactive Glasses for Nonbearing Applications in Total Joint Replacement," *Semin. Arthroplasty*, 17, 102–12 (2007).

[14]Q. Fu, M. N. Rahaman, B. S. Bal, W. Huang, and D. E. Day, "Preparation and Bioactive Characteristics of a Porous 13–93 Glass, and Fabrication into the Articulating Surface of a Proximal Tibia," *J. Biomed. Mater. Res.*, **82A**, 222–9 (2007).

[15]Q. Fu, M. N. Rahaman, B. S. Bal, R. F. Brown, and D. E. Day, "Mechanical and In Vitro Performance of 13-93 Bioactive Glass Scaffolds Prepared by a Polymer Foam Replication Technique," *Acta Biomater.*, **4**, 1854–64 (2008).

[16]Q. Fu, M. N. Rahaman, R. F. Brown, and B. S. Bal, "Preparation and In Vitro Evaluation of Bioactive Glass (13-93) Scaffolds with Oriented Microstructures for Repair and Regeneration of Load-Bearing Bones," *J. Biomed. Mater. Res.*, (2009). In press.

[17]D. E. Day, J. E. White, R. F. Brown, and K. D. McMenamin, "Transformation of Borate Glasses into Biologically Useful Materials," *Glass Technol.*, **44**, 75–81 (2003).

[18]S. D. Conzone and D. E. Day, "Preparation and Properties of Porous Microspheres Made from Borate Glass," *J. Biomed. Mater Res. A*, **88**, 531–42 (2009).

[19]K. P. Fears, Formation of Hollow Hydroxyapatite Microspheres, MS Thesis, Missouri University of Science and Technology, Rolla, MO (2001).

[20]Q. Wang, W. Huang, D. Wang, B. W. Darvell, D. E. Day, and M. N. Rahaman, "Preparation of Hollow Hydroxyapatite Microspheres," *J. Mater. Sci.: Mater. Med.*, **17**, 641-646 (2006

[21]W. Huang, M. N. Rahaman, D. E. Day, and B. A. Miller, "Strength of Hollow Hydroxyapatite Microspheres Prepared by a Glass Conversion Process," *J. Mater. Sci.: Mater. Med.*, **20**, 123-129 (2009).

[22]A. R. Tan, J. M. Barsi, P. S. Jayabalan, M. N. Rahaman, B. S. Bal, G. A. Ateshian, J. L. Cook, and C. T. Hung, "Bioactive Glass (13-93) as a Medium Supplement for Culturing Tissue-Engineered Cartilage," *Transactions of the 55th Annual Meeting of the Orthopaedic Research Society*, Las Vegas, NV, **Vol. 34**, Paper 1314 (2009).

[23]B. S. Bal, M. N. Rahaman, P. Jayabalan, K. Kuroki, M. K. Cockrell, J. Q. Yao, and J. L. Cook, "Comparison of Bone Allograft, Trabecular Tantalum, and Bioactive Glass in Tissue-engineered Osteochondral Grafts," *J. Biomed. Mater. Res. Part B.*, (2009).

[24]M. Brink, T. Turunen, R. Happonen, and A. Yli-Urppo, "Compositional Dependence of Bioactivity of Glasses in the System $Na_2O-K_2O-MgO-CaO-B_2O_3-P_2O_5-SiO_2$," *J. Biomed. Mater. Res.*, **37**, 114–21 (1997).

[25]H. Zhang, I. Hussain, M. Brust, M. F. Butler, S. Rannard, and A. I. Cooper, "Aligned Two- and Three-Dimensional Structures by Directional Freezing of Polymers and Nanoparticles," *Nature Mater.*, **4**, 787–93 (2005).

[26]S. Deville, E. Saiz, and A. Tomsia, "Freeze Casting of Hydroxyapatite Scaffolds for Bone Tissue Engineering," *Biomaterials*, **27**, 5480–9 (2006).

[27]S. Deville, E. Saiz, R. K. Nalla, and A. Tomsia, "Freezing as a Path to Build Complex Composites," *Science*, **311**, 515–8 (2006).

[28]Q. Fu, M. N. Rahaman, F. Dogan, and B. S. Bal, "Freeze Casting of Porous Hydroxyapatite Scaffolds – I. Processing and General Microstructure," *J. Biomed. Mater. Res. Part B: Appl. Biomater.*, **86B**, 125–35 (2008).

[29]J. O. Hollinger and K. Leong, "Poly(α-hydroxy acids): Carriers for Bone Morphogenetic Proteins," *Biomaterials*, **17**, 187–94 (1996).

[30]Y. H. Hu, D. W. Grainger, S. R. Winn, and J. O. Hollinger, "Fabrication of Poly(a-hydroxy acid) Foam Scaffolds Using Multiple Solvent Systems," *J. Biomed. Mater. Res.*, **59**, 563–72 (2002).

[31]A. Sabokar, P. J. Millett, B. Myer, and N. Rushton, "A Rapid, Quantitative Assay for Measuring Alkaline Phosphatase in Osteoblastic Cells In Vitro," *Bone Miner.*, **27**, 57–67 (1994).

[32]Y. Kato, A. Boskey, L. Spevak, M. Dallas, M. Hori, and L. F. Bonewald, "Establishment of an Osteoid Preosteocyte-like Cell MLO-A5 that Spontaneously Mineralize in Culture," *J. Bone Miner. Res.*, **16**, 1622–33 (2001).

[33]J. E. Gough, J. R. Jones, and L. L. Hench, "Nodule Formation and Mineralization of Human Primary Osteoblasts Cultured on a Porous Bioactive Glass Scaffold," *Biomaterials*, **22**, 2039–46 (2004).

COMPARISON OF REACTIONS OF BIOACTIVE GLASSES IN DIFFERENT AQUEOUS SOLUTIONS

Susanne Fagerlund, Leena Hupa, and Mikko Hupa
Process Chemistry Centre, Åbo Akademi University
Turku, Finland.

ABSTRACT

In vitro reactions of bioactive glasses are commonly studied in different types of aqueous solutions, such as simulated body fluid (SBF), TRIS-buffer, and phosphate-buffered saline (PBS). SBF and TRIS have been used to study the reactions of bioactive glasses while PBS has been used to evaluate hydrolysis of polymer/bioactive glass composites. We compared the reactions of 45S5, S53P4, 1-98, and 13-93 in these solutions. Glasses were tested as granulates or fibers. The pH changes were recorded as a function of immersion time. Ions dissolved from the glasses were determined using ICP-AES. Formation of silica and apatite layers on the glasses was characterized using SEM-EDXA. The glasses reacted very differently in the three solutions. SBF produced the expected silica and apatite layers associated with the bioactive behavior. In TRIS no apatite layer formation was observed with the fiber samples. In PBS the apatite layer formation was the most rapid, and the glass itself dissolved at much higher rate than in the other solutions.

INTRODUCTION

In vitro behavior and overall chemical durability of biomaterials are one of the key parameters when assessing novel implant materials. When new glass compositions are studied the main interest often lies in the apatite formation, which is considered to correlate with the bioactivity of the material in physiological solutions[1]. Thus, the in vitro conditions should resemble the in vivo conditions as closely as possible[2]. In vitro reactions, and especially the reaction kinetics, are affected by a large number of experimental variables such as pH and concentration of ions in the solution, circulation of the solution, geometry and texture of the sample, and chemical composition of the glass[3]. Usually, the in vitro experiments are carried out in buffered solutions in order to control the pH changes in the solutions, thus better mimicking the in vivo environments. When a buffer solution is chosen for in vitro experiments the objective and the nature of the experiments need to be taken closely into consideration. In the physiological pH range (6.8-7.4), the ion exchange between the alkali and alkaline earth ions in the glass and H^+ in the solution has been shown to be the dominant reaction. The reaction starts immediately after the glass comes in contact with the aqueous solutions[3, 4]. When the pH increases above 9, silica dissolution, i.e. network dissolution becomes the dominant reaction[5]. Moreover, the pH influences the growth rate of calcium phosphate (Ca-P) phases because the solubility product of Ca in an aqueous solution is pH dependent[6, 7]. In physiological solutions precipitation of hydroxyapatite (HA) is considered as the most thermodynamically stable phase, while in aqueous solutions also other compounds of calcium and phosphate have been observed[8].

In vitro reactions of bioactive glasses are commonly studied in solutions such as TRIS-buffer and simulated body fluid (SBF)[9,10]. The former is pure water buffered with tris-hydroxymethylaminomethane, while the SBF solution resembles the inorganic part of human extracellular fluid. Also different variants of SBF and TRIS, some also containing organic compounds, have been developed. The use of SBF for bioactivity testing has significantly increased after the development of the solution[11]. According to SciFinder Scholar library program, more than 300 scientific papers dealing with SBF and glass have been published by April 2009. Also TRIS is frequently used, mainly because the dissolution reactions of glass are easier to follow in this simpler solution[4, 12]. Weight loss and dissolution of silica from glass samples immersed in TRIS and SBF have been reported to be very similar[2, 10]. However, formation of the calcium phosphate –rich layer on the glass in TRIS depends on the oxide composition of the glass; only if the concentrations of calcium or phosphate ions dissolved from the glass are high enough, hydroxyapatite precipitates.

Sodium phosphate-buffered saline (Na-PBS) is not normally used for in vitro testing of glasses. It has been used mainly to evaluate hydrolysis of polymer/bioactive glass composites[13-16]. In Na-PBS, a distinct HA layer on glass has been reported, but the glass reactions have not been discussed in detail. The Na-PBS is prepared by dissolving sodium chloride, monosodium phosphate, and its conjugate base disodium phosphate into ultra pure water to generate pH values around 7 (pKa=7.21). In comparison, the effective buffering range of Tris (pKa=8.06 at 25°C) is between 7.0 and 9.2. PBS is easily prepared and the pH of the solution is not sensitive to temperature changes in contrary to TRIS-buffered solutions. The high initial phosphate concentration of PBS gives a fast precipitation when calcium species are added to the solution. Thus, PBS is not an appropriate choice for in vitro testing of lime containing bioactive glasses.

In this work we compared the in vitro reactions of four bioactive glasses in TRIS, SBF, and Na-PBS. The in vitro reactivity was measured as changes in the ion concentrations and pH of the immersion solutions as functions of the immersion time. Also formation of reaction layers on the glasses was characterized. The goal was to establish the suitability of these buffered solutions for in vitro testing of glasses and composites containing glasses as one constituent. The trends observed in the reactivity of the glasses in the different solutions can be used e.g. when comparing experimental results in one of the solutions with reported observations of the same glass in the other solutions.

EXPERIMENTAL

Three different buffered solutions: TRIS, SBF, and PBS were used to compare the in vitro reactions of bioactive glasses 45S5, S53P4, 13-93, and 1-98. Glasses 45S5 and S53P4 are commercially available glasses used in clinical applications, while the bioactivity of 13-93 and 1-98 has been tested in vivo[17, 18]. In addition, ultra pure water (Millipore) was used to study the pH behavior of samples in an unbuffered environment. The ion concentrations of the solutions are given in Table I and the glass compositions in Table II. The initial pH of the immersion solutions was adjusted to 7.34 (37°C).

Table I. The ion concentrations of the immersion solutions (mmol/l).

Ion	Blood plasma	SBF	Na-PBS	Tris
Na^+	142	142	156.2	-
K^+	5	5	-	-
Mg^{2+}	1.5	1.5	-	-
Ca^{2+}	2.5	2.5	-	-
Cl^-	103	147.8	100.9	45
HCO_3^-	27	4.2	-	-
HPO_4^{2-}	1	1	24.9	-
SO_4^{2-}	0.5	0.5	-	-
$H_2PO_4^-$	-	-	5.5	-

Glasses 45S5 and S53P4 were tested as granulates (500-800 µm). The glass melting and crushing procedures have been described in detail elsewhere[19]. The mass of granulates was 1.5g

(100 mg glass/ml solution)[19]. Glasses 13-93 and 1-98 were used as fibers (d_{fiber} 20-140 µm). Both 13-93 and 1-98 are compositions which allow drawing of continuous fibers[20, 21]. PLA coated 13-93 fibers were kindly provided by Vivoxid Ltd. The 1-98 fibers were downdrawn semiautomatically from glass melt in a laboratory fiber drawing equipment[21]. The variation of fiber diameter (specified with SEM) between different batches was large. Thus, instead of studying the fibers at a constant SA/V – ratio, a constant mass, 15 mg of fibers per 16 ml of solution was used (SA/V≈0.4 cm^{-1}). The continuous fibers were cut into 2-3 cm pieces, and placed freely in a polystyrene container.

In vitro studies were performed by soaking the samples in the solutions for 1 hour to 14 days. All immersion tests were performed in a water bath at 37°C without agitation, and the solutions were not changed during the immersion. The pH of the samples was measured periodically during the immersion (Mettler Toledo). After soaking, the solution and the sample was separated. The samples were rinsed three times with ultra pure water and ethanol to remove residual salts, and finally preserved in ethanol. The final pH of all the immersion solutions was recorded after the separation. The pH results presented are mean values of three to six parallel samples. In general, the pH varied only slightly (<0.05 pH units) between the parallel samples. Error bars are given as minimum and maximum values for time points with larger variations.

Ion concentrations of the solutions were measured with an inductively coupled plasma atomic emission spectrometry (ICP-AES; Thermo Jarrell Ash AtomScan^{TM} 25). The samples were diluted prior to the analysis. The analyzed elements and the corresponding emission lines (five replicates per line) were silicon (λ=251.611 nm), sodium (λ=589.592 nm), potassium (λ=766.490 nm), calcium (λ=317.933 nm), magnesium (λ=279.553 nm), and phosphorus (λ=213.617 nm). Calibration standards were prepared with 1000 ppm solutions (UltraGrade^{TM}, Ultra Scientific), and a multipoint calibration procedure was used (0-50 ppm). Standard sample concentrations were measured periodically to ensure the accuracy and stability of the calibration. ICP-AES results in this paper give an average of three samples. As the initial solutions have highly different initial ion compositions the concentrations are given as:

$$\Delta c = c_i - c_o \quad (1)$$

where c_0 is the initial value in the solution and c_i is the elemental concentration after i hours of immersion.

Formation of different reaction layers was monitored with a scanning electron microscope equipped with energy dispersive X-ray analysis (FEG-SEM, LEO 1530 from Zeiss/EDXA from Vantage by Thermo Electron Corporation). Prior to the analysis the samples were dried at 50°C, embedded into epoxy, cut and polished in order to get the cross-sections. Surface analysis of some of the fibers was also performed.

Table II. Chemical composition of the experimental glasses (wt-%).

Glass	Na_2O	K_2O	MgO	CaO	P_2O_5	B_2O_3	SiO_2
1-98	6.0	11.0	5.0	22.0	2.0	1.0	53.0
13-93	6.0	12.0	5.0	20.0	4.0	-	53.0
45S5 (Bioglass®)	24.5	-	-	24.5	6.0	-	45.0
S53P4	23.0	-	-	20.0	4.0	-	53.0

RESULTS AND DISCUSSION

The pH changes of the immersion solutions were followed as a function of immersion time. The pH increase in unbuffered medium, ultra pure water, is shown in Figure 1. The pH increases rapidly to very high pH (>9) with all the samples, except 13-93, which shows slower increase. A rapid pH increase in water has been reported earlier by many research groups[4, 19, 22]. The high pH values increases the network dissolution as discussed earlier. The lower values of 13-93 were expected because of the PLA coating that protects the fibers[20]. As the fibers were tested as short filaments a large surface area of 13-93 fibers was still available to immediately react with the solution.

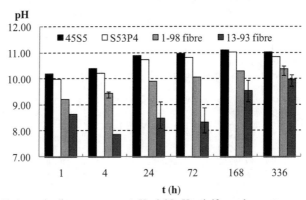

Figure 1. The pH change in ultra pure water, ΔpH <0.05 pH unit if error bars not presented.

Figures 2 and 3 show the pH changes as functions of time for the buffered solutions. The differences in buffering capacity of the solutions can be seen clearly. The pH trends of solutions containing 45S5 and S53P4 granulates are similar (Figure 2). These glasses undergo a rapid ion exchange reaction, and already after four hours of immersion the effective buffering range of Na-PBS has been passed. After this point the pH continues to increase to similar levels than with the unbuffered solution, water. The fast reactions of 45S5 and S53P4 in contact with aqueous solutions are well known. S53P4 has been shown to have lower reaction kinetics than 45S5[23]. The pH increase in SBF and TRIS follows the same trends as reported for these glasses[2, 22].

The overall increase of pH was lower with the fiber samples than with granulates; this was also expected because of two reasons: the different SA/V relationship, and the different reported reactivity of the glass compositions[24]. The differences in pH in the different solutions are not as pronounced for fiber samples as for the granulates (Figure 3). Even though, small differences can be observed; the highest pH increase is observed in Na-PBS and the lowest in SBF. Both fiber compositions follow the same trends, even though the actual values of 1-98 and 13-93 cannot be directly compared because of the 13-93 fibers were coated.

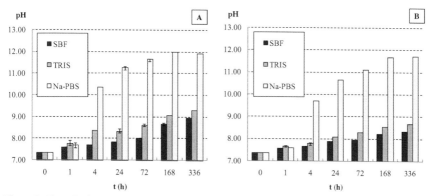

Figure 2. The pH change in SBF, TRIS, and Na-PBS, ΔpH <0.05 pH unit if error bars not presented. (A: 45S5, B: S53P4).

The pH results indicate that both with granulates and fibers SBF had the best buffering capacity and Na-PBS the lowest. The difference in the buffering capacities between SBF and TRIS is depends on the complexity of SBF. The list of possible chemical reactions taking place in SBF is long including e.g. phosphate and carbonate reactions capable of buffering the pH[8]. Recently, the buffering capacity of SBF has been tested by titrating SBF with NaOH[25]. When comparing the titration curve with the calculated curve of TRIS, SBF was found to have a better buffering capacity.

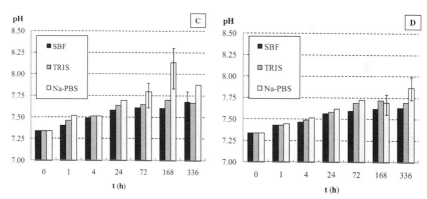

Figure 3. The pH change in SBF, TRIS, and Na-PBS, ΔpH <0.05 pH unit if error bars not presented (C: 1-98, D: 13-93)

Figure 4 shows the changes in the ion concentrations of Si, Mg, K, Ca, and P in the different solutions at different time points for 13-93 fibers. It has been observed that the silicon release rate is higher than or equal to the rate of Si-rich layer formation during the first hours of immersion[26]. The results obtained in this study are in agreement with this observation. The silicon release after one hour was fastest in Na-PBS, but already after four hours of immersion the release rate was lowest, indicating a formation of a diffusion barrier on the fiber surface. In SBF the silicon release during the first 8 hours was found to be faster than in TRIS, but after 24 hours of immersion the silicon concentration approached a constant value in both solutions (SBF 65, TRIS 70 ppm at 120 h) with an apparent trend to saturation. Similar silicon release behavior has been obtained in other studies where SBF and TRIS have been compared[2, 6]. The final measured silicon concentration in Na-PBS was clearly lower (50 ppm).

After the first hour of immersion the change in the magnesium concentration was at the same level in all the solutions. In TRIS the overall magnesium release rate was the highest, and it reached a value of 20 ppm at 120 hours. The release rate was though found to decrease after eight hours of immersion indicating that the growing Si-layer retards the mass transport to the solution. In SBF the increase in magnesium concentration was more moderate. This can partly depend on that the initial concentration in SBF (36.5 ppm) gives a lower concentration gradient than in TRIS. As with the silicon release rate also the magnesium release rate from the fibers was found to be clearly lowest in Na-PBS. In Na-PBS the magnesium release decreased after 72 hours which was not observed in the other two solutions. Though, the standard deviations of these samples were relatively large.

The potassium ion concentration increased fast during the first hours. When analyzing low concentrations of potassium with ICP-AES, the uncertainty of the analysis increases because the sensitivity for potassium is low. Thus, some caution is needed when interpretating these results is done. In Na-PBS the potassium concentration reaches a constant value (32 ppm) after 24 h of immersion. In TRIS the concentration is increasing still at 120 h, and is double compared to that of Na-PBS. The concentration increase in SBF was similar to TRIS. The sodium release was only followed in TRIS. The sodium concentration followed a similar trend as described with potassium. At 24 hours the sodium level was 22 ppm and 30 ppm at 120h. The sodium and potassium in the solutions were correlated with the ration in the glass.

In general, the decrease of calcium and phosphorous in the solution is correlated to the formation of the Ca-P layer. The concentrations of calcium and phosphorus was found to increase in SBF up to 8 hours of immersion, and after which the concentrations decreased, indicating that the layer formation was faster than the leaching after this time point. TRIS is reported to form a soluble complex with calcium ions which might interfere with the Ca-P layer formation[6]. In TRIS the calcium release trend was similar to magnesium, and the final measured value was 95 ppm. The phosphorus concentration was also found to increase in TRIS. Because of the low amount of P_2O_5 in 13-93 the measured values were also low, around 5 ppm after 120 h. Thus no or only sporadic Ca-P layer formation on 13-93 is expected in TRIS.

As the initial concentration of phosphorus in Na-PBS is high the calcium content in the solution will be the limiting factor for layer formation. The release trend of calcium is similar to SBF, but the released amount is much smaller. This suggests that the Ca-P layer formation starts directly after suitable nucleation sites are available. After 48 hours the calcium content was found to reach a constant value around two ppm. The phosphorus concentration follows quite closely the trend of calcium. Even though, the phosphorus increase observed in Na-PBS during the first hours is most likely to be attributed to measurement accuracy than to actual increase. The negative values indicate that the phosphorus is consumed in layer formation.

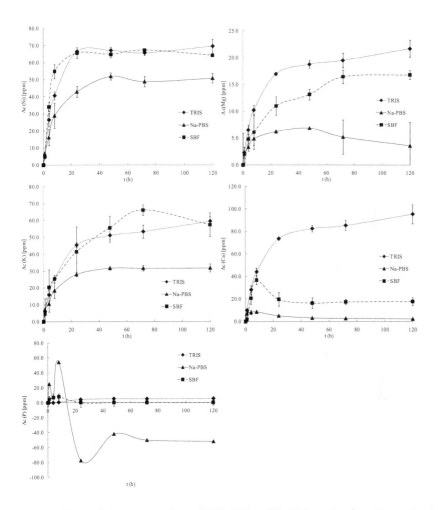

Figure 4. Changes in ion concentrations of TRIS, SBF, and Na-PBS as a function of immersion time of 13-93 fibers.

The ion release from 45S5 and S53P4 granulates at some time points are presented in Table III and IV. The ion release trends in TRIS and SBF are very similar for both glasses. As expected, somewhat higher concentrations were measured in TRIS.

The time interval for phosphorus decrease in the solutions with 45S5 is similar to the observations reported in literature[23, 27]. In Na-PBS, calcium and phosphorus followed the trends described for fiber samples, but silicon was found to increase with time to very high levels (after 24 h: 45S5 ~200 ppm, and S53P4 ~400 ppm). This is attributed to the high pH of the solution (>9), which increases the silica solubility. The clear increase in silicon concentration between four and six hours of immersion of 45S5 granulates in Na-PBS suggest that the buffering capacity of the solution has been surpassed. The high pH has also been found to enhance the Ca-P precipitation because the supersaturation level changes as a function of pH[6]. The high release of silica indicates network dissolution. The intensive network dissolution simultaneously with precipitation of Ca-P is likely to rapidly decrease the mechanical properties of the glasses in Na-PBS.

Table III. Changes in ion concentrations of the immersion solutions as a function of time for 45S5.

	Ac ± stdev (mg/l)									
	TRIS				SBF				Na-PBS	
Time (h)	Na	Ca	Si	P	Na	Ca	Si	P	Ca	Si
0	0	0.0	0.0	0.0	0 ± 18	0 ± 1	0 ± 0	0 ± 1	0	0
1	69	41.0	17.5	6.3	66	35	19	0	13	17
4	91 ± 1	53 ± 3	23.1 ± 0.4	2.3 ± 0.8	173	43	22	-3	8	24
6	105	56	26.2	3.0	61	39	21	-5	3	171
24	176 ± 1	93 ± 14	37 ± 2	3.2 ± 2.4	124 ± 69	72 ± 2	34 ± 1	-11 ± 1	0.2 ± 0.3	237 ± 82
144	435 ± 2	233 ± 10	70 ± 2	3.6 ± 3.7	116 ± 65	115 ± 3	52.7 ± 0.4	< DL ± 5	0.0 ± 0.0	952 ± 38

Table IV. Changes in ion concentrations of the immersion solutions as a function of time for S53P4.

	Ac ± stdev (mg/l)									
	TRIS				SBF				Na-PBS	
Time (h)	Na	Ca	Si	P	Na	Ca	Si	P	Ca	Si
0	0.0	0	0	0.0	0	0	0	0	0.0	0
1	34	18	9	0.6	30 ± 94	4.9 ± 0.2	11 ± 2	-4 ± 2	9.0 ± 0.2	13 ± 4
4	50 ± 3	22 ± 3	13 ± 1	0.6 ± 0.2	37 ± 6	16 ± 4	17 ± 3	-6.2 ± 0.5	7.7 ± 0.7	21.2 ± 0.2
24	124	65	27	0.9	127	43	26	-11	1.4	404.6

Layer formation on 13-93 fibers was measured at the surface and cross-section. Clear differences in the different solutions in the layer formation were detected. The time scale of layer formation differed, and the composition of the formed layers showed differences. Figure 5 shows changes given by EDXA in the oxide composition of the surface layer on 13-93 fiber as a function of immersion time in TRIS, SBF, and Na-PBS.

In TRIS the first change, the formation of a Si-rich layer, was observed after four hours of immersion. In general, the surface in TRIS became more depleted in other components with time and only on some fibers HA precipitation was observed. The surface observations were verified by cross-sectional analysis.

Figure 6 shows that only Si-rich layer thickness increased with time. The thickness of the reaction layers at the same time point varied between individual fibers and samples. This was most likely due to differences in the initial diameter and/or local saturations in the solution causing easier precipitation. Similar uneven layer formation has also been reported earlier[21]. Thus, the layer thicknesses at different time points were not compared to each other. Even though, a clear trend towards increasing layer thickness as a function of time was noticed.

In SBF already after one hour an increase in silica content was noticed, and after eight hours the surface is completely covered by a Si-rich layer (Fig. 5 and 6). At 72 h a distinct Ca-P layer is formed. The cross-section shows three different layers: Si-rich, mixed Si+Ca-P, and Ca-P layer. Separation into the different layers became pronounced with increasing time (Figure 6). At120 h some areas show higher Ca-P ratio than given for HA. 13-93 fibers have been reported to form Si-rich layer with Ca-P precipitation on their surface after 24 hours in SBF[28].

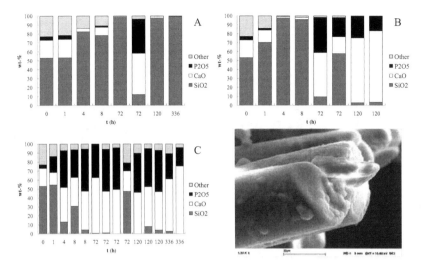

Figure 5. EDXA of the oxide composition on the 13-93 fiber surfaces. (A: TRIS, B: SBF, and C: Na-PBS). The SEM-image (1.5kx) shows the surface after 120 h in Na-PBS.

13-93 fibers reacted fastest in Na-PBS. After one hour of immersion the amount of phosphorus had increased whereas the calcium content had decreased. These results confirm the ICP data. After four hours the surface was covered with a Ca-P layer, and also other elements were found incorporated into the structure. With time only the amount of Ca and P increased on the surface, and the structure became more like the typical cauliflower HA. The SEM image in Figure 5 shows the surface of 13-93 after 120 hours in Na-PBS. The entire surface is covered with Ca-P. The Ca/P ratio on the surface varied with time: at short immersion times the ratio was close to 1.2 while after 72 h the ratio varied between 1.2 and 4. This suggests that also other calcium compounds than HA precipitated.

In Na-PBS a clear Ca-P layer formation was found on top of the Si-rich layer (Fig. 6). A very thin mixed layer can be observed, but it is clearly thinner than formed in SBF. As a function of time the reaction layer thickness clearly increased. EDXA showed the same trend for magnesium as observed with ICP-AES. Magnesium first leached from the fiber surface during the first hours of immersion, but again increased after 8 hours on some of the fibers. Most likely the magnesium was incorporated into the Ca-P layer. Glasses containing magnesium have been reported to form magnesium substituted apatites on the silica layer[29]. Precipitation of magnesium was not found in all the analyzed surfaces, thus explaining the high standard deviations obtained in the ICP analysis of Na-PBS solution. Similar magnesium behavior was not observed in the other solutions. The analysis of 1-98 fibers indicated that the reaction trends were highly similar to those of 13-93 in all the solutions, and thus the results are not presented here.

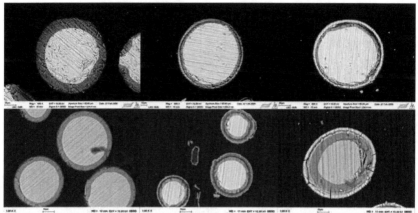

Figure 6. SEM images on the cross-sectional surfaces of 13-93 fibers after 72 h (above, magnification 500x) and 2 weeks (below, 1kx) in TRIS (left), SBF (middle), Na-PBS (right).

Only the cross-sections of 45S5 and S53P4 granulates at 72 hour were analyzed. The differences between the layers formed in the three solutions were not as large as with fiber samples. Both Si-rich as well as Ca-P-rich layers were formed in all solutions, but their minor element content varied. In TRIS and SBF the 45S5 and S53P4 samples formed very similar layers. For both glasses three layer structures was indicated with EDXA. Thus, if the initial leaching of calcium and phosphorus is sufficient, layer formation in TRIS takes place. The compact granulate bed in the static solutions also caused large differences between the individual particles.

The layers formed in Na-PBS seemed to be thicker and more even than in the two other solutions. Even in Na-PBS three layers were observed with both on both glasses, but the elemental compositions of the layers was ambiguous. In all layers all elements present in the glasses could be identified, but in different ratios. In both glasses the outermost layers were enriched by Ca and P, and the layer next to the unreacted glass was Si-rich. The results support the ICP results. Because of the extremely high initial concentrations of sodium in PBS, the outer layers are not depleted in sodium.

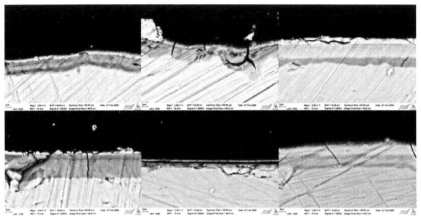

Figure 7. SEM images (2.5kx, the bar 2μm) on the cross-sectional surface of 45S5 (above) and S53P4 granulates after 72 h in TRIS (left), SBF (middle), Na-PBS (right).

CONCLUSION

The in vitro reactions of four bioactive glasses in TRIS, SBF, and Na-PBS were studied. 45S5 and S53P4 were studied in identical conditions as granulates. The buffering capacity of Na-PBS was limited, and the pH increased more than in the two other solutions. Partly this gave increased glass dissolution, but also faster Ca-P formation. SBF was found to have a slightly better buffering capacity than TRIS. In TRIS and SBF a three layered structure consisting of Si-rich, mixed, Ca-P formed, thus confirming the bioactivity of the glasses. Both solutions are well suited for assessing the layer formation on glasses, but when studying the dissolution kinetics via ion concentration determination TRIS would be a preferable choice.

13-93 and 1-98 were studied as fibers. In Na-PBS very fast layer formation was observed in contrary to the other two solutions. At the surface area volume ratio used the pH change was highest in Na-PBS. In TRIS no or only sporadic Ca-P layer formation was noticed. However, in SBF three distinct layers were found. For glasses not releasing sufficient amount of Ca and P for Ca-P precipitation the TRIS might fail to show the possible bioactivity of the glasses.

This study shows that choice of immersion solution and conditions is highly important. Depending on experimental conditions Na-PBS, leads to faster glass dissolution or Ca-P layer formation due to the limited buffering capacity and initial phosphorus content in the solution, and thereby can give misleading information on the bioactivity of the glass. TRIS would be ideal for studying dissolution behavior of glasses in physiological pH ranges.

ACKNOWLEDGEMENT

The National Centre of Excellence Programme by the Academy of Finland and the Graduate School of Chemical Engineering are acknowledged for financial support. Vivoxid Ltd. is acknowledged for the collaboration. Ms Na Li is thanked for help with sample preparation and Mr Linus Silvander for SEM-EDX analysis.

REFERENCES
[1]L.L. Hench, Bioceramics: From Concept to Clinic, *J. Am. Ceram. Soc.*, **74**, 1487-510 (1991).

[2]Ö.H. Andersson and K.H. Karlsson, Corrosion of bioactive glass under various in vitro conditions, In: Heimke G, Soltesz U, Lee AJC, *Clinical implant materials, advances in biomaterials*, Elsevier, 259-64, (1990).

[3]A.Paul, *Chemistry of glasses*, Chapman and Hall, London, (1982).

[4]M. Cerruti, D. Greenspan and K. Powers, Effect of pH and ionic strength on the reactivity of Bioglass® 45S5, *Biomaterials*, **26**, 1665-74 (2005).

[5]T.M. El-Shamy, J. Lewins and R.W. Douglas, The dependence on the pH of the decomposition of glasses by aqueous solutions, *Glass technology*, **13**, 81-87 (1972).

[6] J. Hlavac, D. Rohanova and A. Helebrant, The effect of TRIS-buffer on the leaching behaviour of bioactive glass-ceramics, *Ceramics - Silikáty*, **38**, 119-22 (1994).

[7]K.H. Karlsson, R. Backman and M. Hupa, An equilibrium study of phosphate precipitation on bioactive glass, *Key Eng Mat*, **218-220**, 103-8 (2002).

[8]X. Lu and Y. Leng, Theoretical analysis of calcium phosphate precipitation in simulated body fluid, *Biomaterials*, **26**, 1097-108 (2005).

[9]R.A. Durst and B.R. Staples, Tris/Tris·Cl: A Standard Buffer for Use in the Physiologic pH Range, *Clin. Chem.*, **18**, 206-8 (1972).

[10]T. Kokubo, H. Kushitani and S. Sakka, Solutions able to reproduce in vivo surface-structure changes in bioactive glass-ceramics A-W, *J Biomed Mater Res*, **24**, 721-34 (1990).

[11]M. Bohner and J. Lemaitre, Can bioactivity be tested in vitro with SBF solution?, *Biomaterials*, **30**, 2175-9 (2009).

[12]M.G. Cerruti, D. Greenspan and K. Powers, An analytical model for the dissolution of different particle size samples of Bioglass® in TRIS-buffered solution, *Biomaterials*, **26**, 4903-11 (2005).

[13]H. Yun, S. Kim and Y. Hyun, Preparation of bioactive glass ceramic beads with hierarchical pore structure using polymer self-assembly technique, *Mater. Chem. Phys.*, **115**, 670-6 (2009).

[14]M. Kellomäki, H. Niiranen, K. Puumanen, N. Ashammakhi, T. Waris and P. Törmälä, Bioabsorbable scaffolds for guided bone regeneration and generation, *Biomaterials*, **21**, 2495-505 (2000).

[15]T. Niemelä, H. Niiranen, M. Kellomäki and P. Törmälä, Self-reinforced composites of bioabsorbable polymer and bioactive glass with different bioactive glass contents. Part I: Initial mechanical properties and bioactivity, *Acta Biomater.*, **1**, 235-42 (2005).

[16]T. Niemelä, H. Niiranen and M. Kellomäki, Self-reinforced composites of bioabsorbable polymer and bioactive glass with different bioactive glass contents. Part II: In vitro degradation, *Acta Biomater.*, **4**, 156-64 (2008).

[17]M. Brink, The Influence of Alkali and Alkaline Earths on the Working Range for Bioactive Glasses, *J Biomed Mater Res*, **36**, 109-17 (1997).

[18]A. Itälä, J. Koort, H. Ylänen, M. and H. Aro, Biologic significance of surface microroughing in bone incorporation of porous bioactive glass implants, *J Biomed Mater Res*, **67A**, 496-503 (2003).

[19]D. Zhang, M. Hupa and L. Hupa, In situ pH within particle beds of bioactive glasses, *Acta Biomater.*, **4**, 1498-505 (2008).

[20]E. Pirhonen and P. Törmälä, Coating of bioactive glass 13-93 fibres with biomedical polymers, *J. of Mat. Sci*, **41**, 2031-6 (2006).

[21]H. Arstila, M. Tukiainen, L. Hupa, H. Ylänen, M. Kellomäki and M. Hupa, In vitro reactivity of bioactive glass fibers, *Advances in Science and Technology*, **49**, 246-51 (2006).

[22]D. Zhang, M. Hupa, H.T. Aro and L. Hupa, Influence of fluid circulation on in vitro reactivity of bioactive glass particles, *Mater. Chem. Phys.*, **111**, 497-502 (2008).

[23]L.L. Hench, Ö.H. Andersson and G.P. LaTorre, The kinetics of bioactive ceramics part III : surface reactions of bioactive glasses compared with an inactive glass, *Bioceramics*, **4**, (1991).

[24]D. Zhang, E. Vedel, L. Hupa, H. Aro and M. Hupa. "Predicting Physical and Chemical Properties of Bioactive Glasses from Chemical Composition. Part III, Chemical Composition and In *vitro* Reactivity of Glasses, *Glass Technol. Eur. J. Glass Sci. Technol. A*, **50**, 1-8 (2009).

[25]D. Zhang, In vitro characterization of bioactive glass, Academic dissertation, Åbo Akademi, (2008).

[26]Ö.H. Andersson, J. Rosenqvist and K.H. Karlsson, Dissolution, leaching, and Al_2O_3 enrichment at the surface of bioactive glasses studied by solution analysis, *J Biomed Mater Res*, **27**, 941-8 (1993).

[27]C.Y. Kim, A.E. Clark and L.L. Hench, Early stages of calcium-phosphate layer formation in bioglasses, *J. Non Cryst. Solids*, **113**, 195-202 (1989).

[28]E. Pirhonen, H. Niiranen, T. Niemelä, M. Brink, P. Törmälä, Manufacturing, mechanical characterization, and in vitro performance of bioactive glass 13-93 fibers, *J. Biomed. Mater. Res., Part B*, **77B**, 227-33 (2006).

[29]V. Cannillo, F. Pierli, I. Ronchetti, C. Siligardi and D. Zaffe, Chemical durability and microstructural analysis of glasses soaked in water and in biological fluids, *Ceram. Int.*, **In Press, Accepted Manuscript.**

COMPARISON OF SELF-BONDED THREE DIMENSIONAL BIOACTIVE GLASS FIBER SCAFFOLDS AFTER IN-VIVO IMPLANTATION IN RATS

Steven B. Jung
Graduate Center for Materials Research, Materials Science and Engineering Department
Missouri University of Science and Technology
Rolla, MO, 65409-1170

Delbert E. Day
Graduate Center for Materials Research, Materials Science and Engineering Department
Missouri University of Science and Technology
Rolla, MO, 65409-1170

Roger F. Brown
Department of Biological Sciences
Missouri University of Science and Technology
Rolla, MO, 65409-1170

ABSTRACT

Three dimensional assemblies of fibers made from bioactive glass compositions were investigated to determine their usefulness as scaffolds for the repair and regeneration of hard and soft tissues. The microstructure, properties and in-vivo performance were measured for scaffolds (7mm diameter and 2mm thick) made from an interconnected network of discontinuous fibers of bioactive glasses such as 45S5 and 13-93 that were arranged in a random oriented geometry. Scaffolds, made from fibers whose diameter ranged from approximately 100 up to 300 microns, were implanted subcutaneously in laboratory rats for up to four weeks. Scaffolds comprised of randomly oriented fibers seeded with mesenchymal stem cells prior to implantation contained significant new bone-like material after four weeks in-vivo. No bone-like material was observed in un-seeded scaffolds as expected, but these scaffolds were completely penetrated by new soft tissue with noticeable amounts of new vascularization.

Histological staining techniques were used to asses the quality of the infiltrating soft tissue and to identify cells and tissues of interest. Biological staining and scanning electron microscopy back scattered imaging (SEM-BSI) showed new bone tissue had formed throughout the scaffolds seeded with mesenchymal stem cells. Scanning electron microscopy electron dispersive spectroscopy (SEM-EDS) analysis of the cross-section of a 13-93 and 45S5 fiber implanted four weeks was used to measure the compositional change of each fiber type to calcium phosphate, demonstrating the reactivity of both materials *in-vivo*.

INTRODUCTION

Bone density, especially in women, has been documented to decrease with age [1]. Biologically compatible replacement materials of variable strength are needed to deal with this problem [2,3]. Cortical bone is the strongest type of bone, with a compressive strength that ranges from 117 to 167 MPa [2,3]. Cervical vertebrae and lumbar bone, are weaker having a range of compressive strength between 5 to 10 MPa [3], while cancellous bone, a spongy bone material, has a compressive strength of only 0.2 to 4MPa [2,3].

One idea for the regeneration of bone has been to use ceramic materials for bone scaffolds. This option has been widely researched in terms of both materials and processing techniques [1,4-16]. The end result is often a sintered ceramic three-dimensional scaffold with high porosity (70-90%) and relatively low compressive strength (<0.4MPa) [5,7-8]. Ceramics most often studied for bone scaffold use are stoichiometric hydroxyapatite (HA) with the ideal composition of ($Ca_{10}(PO_4)_6(OH)_2$) [5,7,9,11,15], other crystalline ceramics such as $CaSiO_3$ or $CaCO_3$ [3,9,10], bioactive glasses like the silicate based 45S5 or 13-93 glasses [1,7,11,13,14,17], and glass-ceramics like bioactive glass-ceramic apatite wollastonite (A/W) [1]. All of these materials have been shown to promote significant cell attachment and tissue growth.

Glassy materials have unique properties that make them attractive for use as biological implants and for three dimensional porous scaffolds. It has been shown that the chemical composition can greatly affect the rate at which a glass either dissolves in liquids similar to those found in the body [18-19] or reacts to form HA. Homogeneous nucleation and crystal growth, such as that of bioactive glass 45S5, makes it possible to engineer glass-ceramic materials with improved mechanical properties such as bending strength, hardness and elastic modulus when properly heat treated [20-24]. Other bioactive glasses, such as the silicate based 13-93, have compositions that are more resistant to crystallization [13,25]. The decreased tendency for crystallization is due to compositional changes that reduce the nucleation and crystal growth rate, which results in a wider working range. This change increases the ability to pull fibers from a melt, which can be self-bonded to form bioactive glass scaffolds of a reasonable mechanical strength with a significant fraction of interconnected porosity [13,14,25].

The scaffolds investigated in the present work consisted of a network of randomly oriented self-bonded glass fibers. Scaffold designations are with respect to the weight fraction of 13-93 glass fibers in each scaffold, so S100 scaffolds were composed of 100 wt% 13-93 glass fibers and S30 was 30 wt% 13-93 and 70 wt% 45S5. The preparation of the scaffolds, porosity, compression strength, and in vitro cell growth of MC3T3-E1 pre-osteoblast cells have been previously described for the present scaffolds [26-27].

The main objective of the present research was to compare the in-vivo response of the S100 scaffold versus the S30 scaffold by seeding the scaffolds with mesenchymal stem cells prior to subcutaneous implantation. The evaluation of the scaffolds included determining the scaffold's ability to support bone tissue formation and growth from an initial mesenchymal stem cell seeding, a visual examination for breakdown of the bonding between the fibers to assess the structural capabilities in-vivo, and a comparison of reactivity between the different glass fiber types in body fluids. These aforementioned criteria should give useful insight into the suitability of a scaffold composed of randomly oriented bioactive glass fibers for the growth and regeneration of bone.

MATERIALS AND EXPERIMENTS

Scaffold Preparation and Physical Properties
Continuous fibers (100 to 300 µm in diameter) of the 45S5 and 13-93 bioactive glass (Table 1) were hand pulled from a melt and broken into 3mm lengths. Fibers for each scaffold were individually weighed for a total of 70mg, so S100 scaffolds contain 70mg 13-93 fiber and S30 contained 49mg 45S5 and 21mg 13-93 fibers. The bioactive glass fibers were placed in a mold consisting of an alumina substrate and a mullite cylindrical tube with a 7mm inside diameter. The molds were then placed inside a box furnace that was pre-heated to 720°C. After 45 minutes of heating, the molds were removed from the furnace and cooled to room temperature. The self bonded scaffolds were removed from the molds and were ready for use [27].

The S100 and S30 scaffolds were previously measured for compression strength and open porosity [26-27]. The S100 scaffold had a compressive strength of 5.3MPa and an average open porosity of approximately 44%, while the S30 scaffold had an average open porosity of 63% and compressive strengths of approximately 0.6MPa.

Table 1. Compositions of Some Bioactive Glasses in mol% and wt% (parentheses denote wt%)

Material	Na$_2$O	K$_2$O	MgO	CaO	P$_2$O$_5$	SiO$_2$
13-93 Glass	6 (6)	7.9 (12)	7.7 (5)	22.1 (20)	1.7 (4)	54.6 (53)
45S5 Glass	24.4 (24.5)	0 (0)	0 (0)	26.9 (24.5)	2.6 (6)	46.1 (45)
Na$_4$Ca$_4$Si$_6$O$_{18}$	16.7 (17.5)	0 (0)	0 (0)	33.3 (31.6)	0 (0)	50 (50.9)

Scaffold Sterilization and Seeding with Mesenchymal Stem Cells

All scaffolds were washed twice with ethyl alcohol and heat sterilized at 250°C for 2.5 hours in a small box furnace prior to cell seeding or implantation. Half of the scaffolds were seeded with 150,000 mesenchymal stem cells harvested from Fisher 344 rats (Fisher Scientific, St. Louis, MO). The cells were harvested from the femur of previously sacrificed rats and cultured for 24 hours in glass dishes with Sigma α-MEM 0644 media (Sigma Aldrich, St. Louis, MO): (10% fetal calf serum, 1% penicillin/strep antibiotic, 50mg/ml ascorbic acid, and 10^{-8} molar dexamethasone) at 37°C and in 95% air/5% CO$_2$ atmosphere.

Animal Experiments

A total of eight male and female Fisher 344, rats ranging in age from eight to ten weeks, were used to assess the biological response of the S100 and S30 scaffolds after being implanted for three and four weeks. The back of each rat was shaved just prior to surgery, sterilized with iodine, and then washed with 70% ethanol to ensure a clean and sterile working area. Each rat was anesthetized with a mixture of isoflourine and medical grade oxygen. The implanted scaffolds are hereafter referred to in by the following designations; scaffold S100 seeded with mesenchymal stem cells is designated S100+, scaffold S100 without mesenchymal stem cells is designated to as S100, scaffold S30 with mesenchymal stem cells is designated S30+ and scaffold S30 without mesenchymal stem cells is designated S30.

One scaffold of each type, S100, S100+, S30, and S30+ was implanted for a total of four scaffolds per animal for a total of 32 scaffolds. Each scaffold was implanted subcutaneously in a pocket that was formed in the back of each rat, one above each shoulder and one above each hind leg, between the skin and the muscle as shown in Fig. 1. The pockets were approximately 15mm wide and 15mm long to ensure that each scaffold was inserted away from the incision site. The incisions were closed with super glue (Krazy® Glue, Elmers Products inc. Columbus, OH). After implantation, 0.1 mL of Penicillin G Procaine was injected into each thigh of the rat to prevent infection. The rats were placed on a heating pad in a cage with fresh air during recovery.

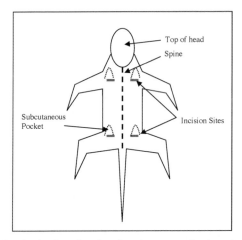

Figure 1. Schematic showing implantation sites for subcutaneous implantation of the S100 and S30 scaffolds in the back of a rat.

Removal of Scaffold from Rats and Scaffold Dehydration

After the allotted three or four week period, the rats were sacrificed in a chamber of CO_2. Each scaffold was surgically removed from the subcutaneous site of each rat and placed in a 10% formalin solution at room temperature (22°C) for 96 hours to fix the tissue. At this point, each scaffold was visually examined as described later.

Each scaffold was dehydrated with a series of seven ethanol solutions ranging from 80 to 100% ethanol using a solution volume ratio of ten times that of the scaffold (~5mL solution/scaffold) [28]. Each scaffold was placed in glass vial and submerged in the ethanol solution. The glass vials were placed in a plastic tray, which was packed in ice to keep the ethanol solution from over heating, microwave irradiated at 80% power at 37°C for two minutes and was continuously submerged in each of the alcohol solutions for a total of fifteen minutes. After dehydrating, each scaffold was placed in xylene for 24 hours to make any tissue present transparent.

Impregnating Scaffolds with Polymethyl Methacrylate (PMMA)

Each scaffold was infiltrated with a series of three solutions for 48 hours each (0%, 50%, and 100% polymerizing agent perkodox mixed with methyl methacrylate monomer (MMA)) to ensure monomer penetration and controlled polymerization. Just prior to infiltration of the 100% MMA solution, the scaffold was attached to a bed of partially polymerized PMMA. The scaffold was then covered with a layer of MMA and slowly polymerized over a four day period at 4°C.

Scaffold Separation, Sectioning, and Polishing

Each scaffold was sectioned at its center, perpendicular to the 7mm diameter using an Isomet[TM] slow speed saw (Model 11-1180, Buehler LTD. Evanston, IL) with a LECO® diamond coated blade (LECO® Corporation, St. Joseph, MI) . The cross section of the scaffold was polished with silicon carbide papers ranging from 320 to 1200 grit. The polished surface was attached to a glass slide with superglue and polished to a final thickness of 50 ±10μm.

Scaffold Cross-section Staining

The 50μm section of each scaffold was stained using Sanderson's[TM] Bone Staining method (Surgipath Canada inc, Winnipeg, Manitoba). The mounted scaffold cross sections were dipped in the preheated stain (50°C) for two minutes, rinsed with distilled water, and blotted dry with a paper towel. The stained sections were then counter stained with a solution consisting of one gram of acid fuchsin, 99mL of distilled water, and one mL acetic acid for 40 seconds at 22°C to differentiate the soft tissue as blue and bone tissue as red. The cell nuclei of both the soft and bone tissue appear as dark blue dots, cytoplasm (a watery substance that fills cells) is light blue, osteoid seam (a mixture of proteins which is secreted by osteoblast which when mineralized becomes bone) is a purple color, bone matrix material is red, and soft tissue is blue. Each stained section was examined at 10X and 40X magnification using an Olympus MI transmitted light microscope (Olympus Corporation, Japan).

SEM Imaging and EDS Compositional Analysis *In-vivo* Scaffolds

Scanning electron microscopy backscattered imaging SEM-BSI, (Hitachi S-570 SEM) was used to determine the compositional change from the center to the outer edge of reacted fibers of 13-93 and crystallized 45S5 fibers (45S5c) that had been four weeks *in-vivo*. Cross sectioned scaffolds were polished with silicon carbide papers to 1200 grit. The scaffolds were sputter coated with gold palladium and attached to aluminum stubs with carbon tape to eliminate charging. For standard-less EDS, there is no set error, as it depends on the elements analyzed and the accelerating voltage used for generation of the EDS spectra. For this EDS analysis, the data plotted is shown to give a general idea of the reaction and leaching of the fibers.

RESULTS AND DISCUSSION

Scaffolds Prior to Implantation (As-Made)

Figure 2 shows a high and low magnification image of the as-made S100 (Figs 2a and b) and S30 (Figs 2c and d) scaffolds prior to implantation. Figures 2a and Fig. 2c show the random orientation of the fibers in the as-made scaffolds. An example of the bonding developed between the fibers after heat treatment is shown in Figs. 2b and 2d. An important difference in the 13-93 and 45S5 fibers after self-bonding is that the 13-93 fibers remain glassy and smooth, whereas the 45S5 fibers crystallize and a rough surface is formed as shown in Fig. 2d.

Figure 2. SEM images of the 13-93 fibers in the S100 scaffold (a), and S30 scaffold (c) after 45 minutes at 720°C. The bonding between 13-93 glass fibers in a S100 scaffold is shown in (b) and a 13-93 fiber bonded to 45S5c fiber in S30 scaffold in (d). [26]

Evaluation of Scaffolds Implanted in Rats
    Upon removal, the first evaluation was a visual assessment of robustness of the scaffold. Figure 3 shows three scaffolds, (a) an as-made S100 scaffold prior to implantation, (b) a S100+ scaffold after four weeks *in-vivo*, and (c) a S30+ scaffold after four weeks *in-vivo*. The diameter of the two implanted scaffolds was essentially unchanged from the as-made control, and both scaffolds had sufficient strength to remain intact after four weeks *in-vivo*.
    Significant soft tissue had infiltrated both scaffolds after four weeks *in-vivo* as is clearly evident from Figs. 3b and 3c. Healthy tissue had grown into the scaffolds, and there was no evidence of infection or inflammation in the surrounding tissue to suggest rejection of the scaffolds. Another interesting observation was that several blood vessels had grown into the S100 and S30 scaffolds; see black circles in Figs. 3b and 3c. The remaining scaffolds, not shown, at three and four weeks, with and without seeded mesenchymal stem cells, all looked similar to those in Fig. 3 in terms of tissue infiltration and the presence of blood vessels.
    The reaction products of these two bioactive glasses may be causing an angiogenic effect, which in this case is positive, as the penetrating blood vessels are providing nourishment to the interior of the scaffolds. Previous work completed by others on the angiogenic effect of 45S5 glass found that there may be a link between bioactive glasses and angiogenesis *in-vitro* [29-30]. No one to date has reported on proangiogenic effects for 13-93 glass, *in-vivo* or *in-vitro*, but this work shows compelling evidence that there may be a positive angiogenic effect *in-vivo*.

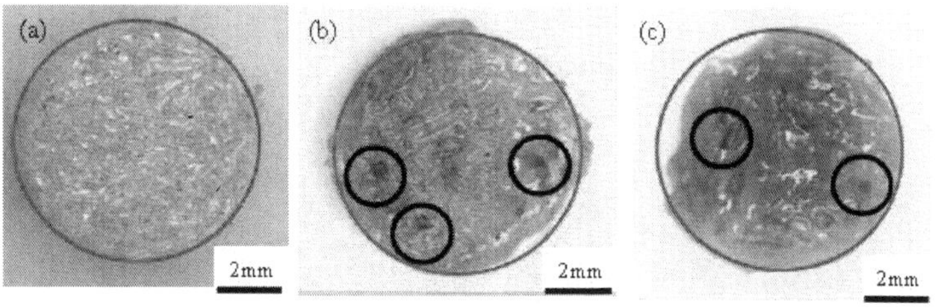

S100 Prior to
Implantation

S100+ after 4
wks *In Vivo*

S30+ after 4
wks *In Vivo*

Figure 3. Appearance of glass fiber scaffolds before and after subcutaneous *in vivo* implantation for four weeks. (a) S100 scaffold prior to implantation, (b) S100+ after four weeks *in- vivo*, (c) S30+ after four weeks *in-vivo*. Red circle is 7mm in diameter and denote the starting diameter of the scaffolds.

Evidence of Bone Formation in Implanted Scaffolds

Regions of red tissue in the histology sections were assumed to be bone tissue, therefore, the results of the scaffold cross sections stained for bone tissue identification in both the S100 and S30 scaffolds are as follows. None of the non-cell seeded scaffolds contained any bone tissue as expected since they had no mesenchymal stem cells seeded on them, and were not adjacent to any natural bone tissue. All of the S100 and S30 scaffolds seeded with mesenchymal stem cells and implanted for three and four weeks had red tissue present after being stained and was assumed to be bone tissue.

The 13-93 fibers in the non-seeded S100 scaffolds in Fig. 4 are the white elliptical regions surrounded by a dark line. None of the non-seeded S100 scaffolds contained any detectable bone tissue (red stained region) after three or four weeks *in-vivo*, as is evident in Figs. 4a and 4b. However, soft tissue had penetrated the outer portion of the unseeded scaffolds and was often present in the center of the scaffold. The presence of this tissue is an indication of the biocompatibility between the soft tissue and the 13-93 fibers.

Three Weeks   S100   Four Weeks

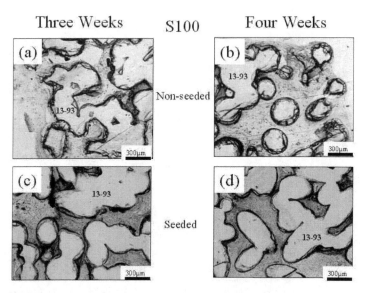

Figure 4. Optical micrograph of randomly selected location in the stained cross section of scaffolds after three or four weeks *in-vivo*. Soft tissue (blue) and bone tissue (red) formation. (a) S100 three weeks, (b) S100 four weeks, (c) S100+ three weeks, (d) S100+ four weeks.

The seeded S100+ scaffolds all contained regions of bone tissue (red regions) after three and four weeks *in-vivo*, as is evident in Figs. 4c and 4d. There is a mix of both soft (blue) and bone (red) tissue present in the three and four week S100+ scaffolds. The amount of bone tissue in the week four scaffold, Fig. 4d, appears to be slightly more than in the three week scaffold, Fig. 4c.

Figure 5 is as enlarged view of Fig. 4d to better show the regions of soft and bone tissue after four weeks *in-vivo*. The blue dots within the blue matrix are nuclei of soft tissue, and the blue dots in the red matrix are nuclei of osteocytes which have incorporated into the bony matrix. In several areas of the scaffold, pockets of bone tissue are attached and growing on the 13-93 fibers. There is a visible reaction layer present on the surface of the 13-93 fibers that formed as the 13-93 fibers reacted with the body fluids of the rat. The reaction layer is the darker gray layer that outlines the 13-93 fibers, and will be described in more detail later.

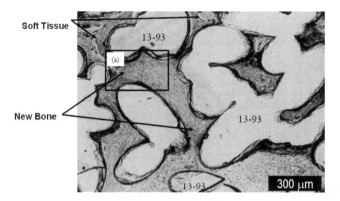

Figure 5. Optical micrograph (100X) of the cross-section of a S100+ scaffold seeded with mesenchymal stem cells after four weeks *in-vivo*. Tissue has been stained where blue denotes soft tissue and red denotes bone (hard) tissue.

The boxed area in Fig. 5 labeled (a) is shown at higher magnification in Fig. 6. The reacted surface layer on the 13-93 fibers varies in thickness between ~10 to 20μm. The dark circles in Fig. 6 encircle osteocytes (blue dots surrounded by a white halo) that are incorporated into the newly formed bone tissue. Only a few examples of the osteocytes in the bony matrix material are shown in Fig. 6, but osteocytes were seen throughout the new bone as shown in Fig. 5. This is important because it shows that the MSC cells have differentiated to become mature bone and that the S100 scaffold is promoting osteogenesis. The osteoid seam (purple tissue) represents un-mineralized secreted proteins of osteocytes and is the growth front for new bone, as pointed out with arrows at both the far right and lower left of Fig. 6.

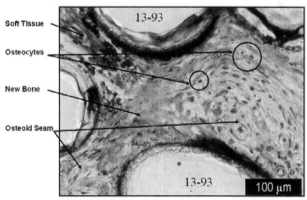

Figure 6. Optical micrograph (400X) of S100+ scaffold cross-section. S100+ scaffold was seeded with mesenchymal stem cells and implanted four weeks subcutaneously in the back of a rat and stained for soft (blue) and bone (red) tissue.

Representative examples of the seeded and non-seeded S30 scaffolds after three of four weeks *in-vivo* are shown in Fig. 7. These cross-sections look significantly different than those for the S100 scaffolds, see Fig. 4. The 45S5c fibers are dark and crystallized as opposed to the glassy white 13-93 fibers. Again, none of the non-seeded scaffolds contained any evidence of bone tissue (red), as is evident from Fig. 7a (S30 three weeks), and Fig. 7b (S30 four weeks). The soft tissue (blue) in Figs. 7a, 7b, and 7c stained relatively light making it difficult to see, however the cell nuclei stained dark blue. The abundance of nuclei through out the cross-sections indicates significant infiltration of soft tissue. Soft tissue (blue) is visible throughout areas of the scaffold cross-section in Fig. 7d, indicating that it had penetrated and survived within the scaffold environment provided by the mixture of 13-93 and 45S5c fibers.

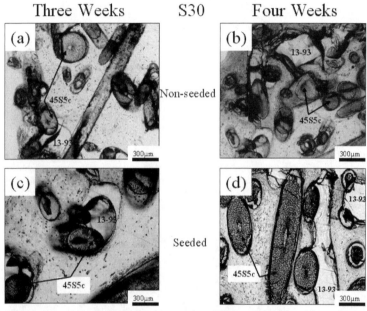

Three Weeks   S30   Four Weeks

Figure 7. Optical micrograph (100X) of the cross-section of S30 scaffolds after three and four weeks *in-vivo* and stained for soft (blue) and bone (red) tissue formation. (a) S30 three weeks, (b) S30 four weeks, (c) S30+ three weeks, (d) S30+ four weeks.

Figures 7c and 7d are representative sections of the S30+ scaffolds in which bone tissue had formed. There is a mixture of both soft and bone tissue present in these scaffolds, and again the total amount of bone tissue is estimated to be slightly higher after four weeks *in-vivo*. In Figure 7d, new bone tissue is seen surrounding both the 13-93 and 45S5c fibers. This suggests that both materials are biocompatible and suitable for bone growth. To the author's knowledge, this is the first time that 13-93 and 45S5c fibers have been placed *in-vivo* in the same site, and from this analysis, both fibers appear to interact positively with the living tissue, and there does not appear to be any significant advantage from a biological point of view between the two fibers.

The surface of the 13-93 and 45S5c fibers differs greatly as the 13-93 fibers are smooth, while the surface of the 45S5c fibers are noticeably rougher as shown in Fig. 2d. The cause of the rough surface is unknown, but one possible explanation is that as the 45S5 glass crystallized and the crystals began to grow together, a liquid or viscous phase surrounding the crystals was pushed out causing the rough surface. The microstructure in Fig. 7 shows bone tissue on both the smooth 13-93 and rough 45S5c surfaces, but a determination of which surface is preferred for tissue attachment and growth cannot be made from the present work.

The enlarged view of Fig. 7d shown in Fig. 8, better shows the seeded S30+ scaffold after four weeks in-vivo. The high level of crystallization that occurred in the 45S5c fibers when they were heated to form a scaffold is evident from the small irregular grains that are present throughout the fibers in Fig. 8. The 45S5c fibers have visible reaction regions surrounding them, light gray to white in color, that is attributed to their being in contact with the body fluids of the rat. Two examples of the bonding between the 45S5c fibers and a 13-93 fiber are shown on the right hand side of Fig. 8. Both fibers labeled as '13-93' in Fig. 8 are attached to two 45S5c fibers. The attachment appears to be from 13-93 fibers softening and sticking to the 45S5c fibers.

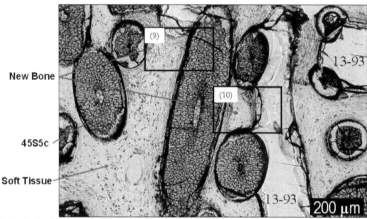

Figure 8. Optical micrograph (100X) of S30+ scaffold cross-section after four weeks in-vivo. The S30+ scaffold was seeded with mesenchymal stem cells and implanted four weeks subcutaneously in the back of a rat and stained for soft (blue) and bone (red) tissue. The darker, textured fibers are crystallized 45S5c fibers, where the glassy 13-93 fibers are transparent and featureless.

An expanded view of the region inside the box labeled (9) in Fig. 8 is shown in Fig. 9. A region of new bone tissue is seen between two 45S5 fibers that are bonded to a 13-93 fiber. Osteocytes are visible in the newly formed bone, as observed in the S100+ scaffolds. New bone tissue appears to be attached to both the 13-93 fiber and the 45S5c fiber, with no apparent preference for either fiber. This indicates that both fibers promote new bone growth. Again, as mentioned about the S100 scaffold, the S30 scaffold has osteocytes embedded throughout the new bone indicating that the MSC cells have differentiated and are forming new bone.

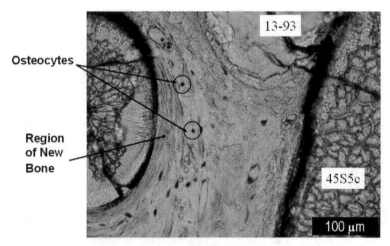

Figure 9. Optical micrograph (400X) of cross-section of the box labeled (9) in Fig. 8. S30+ scaffold was seeded with mesenchymal stem cells and implanted four weeks subcutaneously in the back of a rat and stained for soft (blue) and bone (red) tissue.

The thickness of the reacted region of the 45S5c fiber on the left hand side of Fig. 9 appears to be ~20 to 30μm, which is thicker than the regions seen surrounding the 13-93 fibers. This suggests that the 45S5c fibers react faster than the 13-93 fibers which is consistant with the higher alkali and lower silica content of the 45S5 fibers.

The region, in the box labeled (10), in Fig. 8, is shown as a transmitted light image in Fig. 10a and as a SEM-BSI image in Fig. 10b. Figures 10a and 10b have 45S5c fibers on the left and at the top of the image as denoted by the arrows, and a fiber labeled 13-93 to the right. The X's in Figs. 10a and 10b show regions of the fibers that had reacted in-vivo. These regions are rich in both calcium and phosphorus. The Y's correspond to areas of new bone, stained red in Fig. 10a, and the light gray material attached to the edge of the fibers in Fig. 10b. Soft tissue infiltrated with PMMA, labeled Z, corresponds to the blue tissue in Fig 10a, and the darkest portion of the image in Fig.10b. Figures 10a and 10b look slightly different because the section used for Fig. 10a was stained for histology and the adjacent section of scaffold was prepared for imaging with the SEM. Subsequent polishing removed material from the surface of the scaffold section, which slightly altered the image, but the representative areas of the scaffold are still the same as previously described.

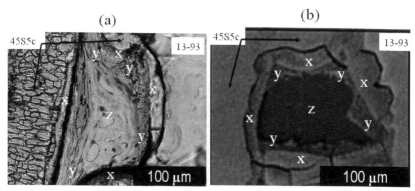

Figure 10. Seeded S30+ scaffold after four weeks *in-vivo*. (a) Optical micrograph (400X) of a portion of the cross-section of a S30+ scaffold. (b) SEM-BSI of same area as shown as (10a). The X's indicate reacted regions on the fibers, the y's indicate areas of new bone, and the Z's are areas of PMMA infiltrated soft tissue.

The new bone tissue found after three weeks *in-vivo* in the seeded scaffolds in the present work is consistent with prior work[10] completed on porous HA and porous calcium carbonate constructs seeded with marrow cells and subcutaneously implanted in rats for two to six weeks. In the prior work[10], three weeks was the minimum time for new bone to become recognizable in the porous HA and porous calcium carbonate constructs.

Analysis of 13-93 and 45S5c Fiber after Four Weeks *In-vivo*

A 13-93 fiber after four weeks *in-vivo* was imaged with SEM-BSI and also measured for compositional change with SEM-EDS. The dotted line in Fig 11a illustrates the boundary between the adjacent new bone and the calcium rich layer of the 13-93 fiber. The new bone is in direct contact with the outer edge of the 13-93 fiber. Figure 11a is an SEM-BSI image of the cross-section of a 13-93 glass fiber that was part of a S100+ scaffold after four weeks. The calculated oxide weight percent, as measured by SEM-EDS, present at the three positions (A,B, and C) shown in Fig. 11a is shown graphically in Fig. 11b. The change in gray scale across the fiber cross-section in Fig. 11a indicates a change in composition from the unreacted glass; position A, to the outer edge at position C. The unreacted portion of the fiber at position A has a composition close to that of starting 13-93 glass (Table 1). Position B is a silica rich layer containing almost 90 weight percent SiO_2, the rest being CaO and P_2O_5. The CaO and P_2O_5 content increased significantly at the outer edge (position C), indicating a calcium phosphate layer had formed. The K_2O, MgO, and Na_2O content decreased to near zero at position C as these constituents are soluble in body fluids and leached from the fibers.

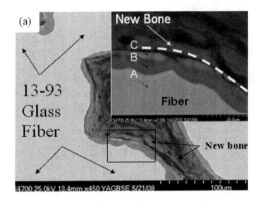

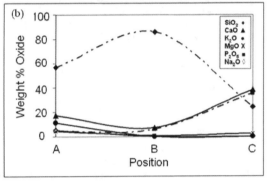

Figure 11. (a) SEM-BSI of a 13-93 glass fiber after four weeks *in-vivo*. (b) Composition (weight % oxide) obtained from SEM-EDS of a reacted 13-93 fiber at the locations A, B, and C. Composition for 13-93 glass given in Table1.

Stoichiometric HA $(Ca_{10}(PO_4)_6(OH)_2)$ has a Ca/P molar ratio of 1.67, and the Ca/P ratio at the outer edge of the 13-93 fiber at position C in Fig. 11b was measured with EDS to be 1.38. This increase in Ca/P ratio at position C indicates that a calcium phosphate material has formed and verifies earlier statements about the formation of a calcium phosphate surface layer on 13-93 fibers due to interactions with body fluids. The formation of a calcium phosphate layer was expected as it has been detected by *in-vitro* testing in SBF [1,11,13, 25, 31] and *in-vivo* in various animal models [1,17].

A 45S5c fiber after four weeks *in-vivo* was imaged with SEM-BSI and also measured for compositional change with SEM-EDS. Figure 12a shows an SEM-BSI image and Fig. 12b shows the corresponding plot of composition across a 45S5c fiber after four weeks *in-vivo*. This fiber has undergone significant reaction after four weeks as is apparent by the thick light gray layer that surrounds an easily distinguishable darker core. The variation in composition across the fiber, as determined by EDS, is shown graphically in Fig. 12b. All of the Na_2O has been leached from the 45S5c fiber, even from the core. This leaching of Na_2O has been previously documented for *in-vitro* conversion of 45S5 glass to HA in dilute phosphate solutions [19].

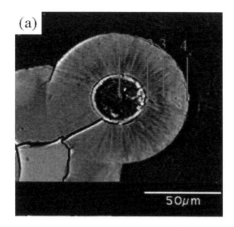

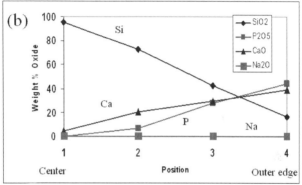

Figure 12. (a) SEMBSI of 45S5c fiber four weeks *in-vivo*. (b) Compositional analysis (weight % oxide) obtained from SEM-EDS of reacted 45S5c fiber as a function of position. Composition of 45S5 glass is shown in Table 1.

The dark core at the center of the fiber in Fig. 12a is greater than 90 weight percent SiO_2. The silica core found in reacted 45S5c fibers in the present work has been similarly described previously with 45S5 glass particles to make calcium phosphate shells *in vitro* [32] and reacting 45S5 glass fibers with phosphate solutions *in-vitro* [18]. The previous report on calcium phosphate shells [32] stated that all of the SiO_2 was eventually released from the 45S5 glass spheres, and lead to a hollow shell of calcium phosphate [32]. It is unclear from the present work if the SiO_2 will be totally dissolved from the center of the reacted 45S5c fibers, but it has been shown that it may, given the right physiological conditions [32].

The calcium phosphate rich layer present at position four of the fiber in Fig. 12a, contains only 17 weight percent SiO_2 while the remaining components are CaO and P_2O_5 at ~40 weight percent each. The Ca/P molar ratio is 1.18 and there are high levels of both CaO and P_2O_5 at position three and four relative to other the other oxides which indicates that a relatively thick (25μm) calcium phosphate layer has formed on the outer edge of the fiber in only four weeks. The formation of the calcium phosphate material at the outer edge of the 45S5c fiber does verify earlier statements about a calcium phosphate reaction region, and should be expected as 45S5 has been documented numerous times to form HA both *in-vitro* with SBF [18,19,23,24], and *in-vivo* in animal models [1,17].

CONCLUSIONS

New bone tissue formed in both the S100 and S30 fiber scaffolds seeded with mesenchymal stem cells at both three and four weeks *in-vivo*. Soft and bony tissues were found in the interior of the cell seeded scaffolds after extraction. Cell seeded scaffold cross-sections showed osteocytes were present in the regions of newly formed bony tissue indicating that the MSC cells were differentiating into mature bone. From this analysis, no preference in fiber composition (45S5c or 13-93) was distinguishable for the soft or bone tissue to grow. Therefore, both materials are excellent for tissue engineering applications.

Several blood vessels which penetrated the scaffolds provided nourishment to the tissues inside both the S100 and S30 of scaffolds. Integration of blood vessels into the interior of the scaffolds could be an indicator that the reaction products from 13-93 and 45S5c fibers could be beneficial in terms of triggering angiogenesis (blood vessel formation) *in-vivo*. Proangiogenic glasses and porous self-bonded bioactive glass scaffolds could be beneficial for new soft tissue repair applications.

Compositional SEM-EDS analysis of 13-93 and 45S5c fibers after four weeks *in-vivo* indicated that calcium phosphate rich layers had formed on the surface of the fibers while other glass components had been resorbed in the body fluids of the rats. The reaction layer thickness present on the 45S5c fibers was in the range of 20 to 30μm while the 13-93 reaction layer was only 10 to 20μm.

Neither scaffold type showed signs of mechanical failure or degradation after four weeks *in-vivo*. Based on these observations and the compressive strength measured previously, both S100 and S30 scaffolds are considered sufficiently strong for *in-vivo* use for cancellous bone applications or for soft tissue. Therefore, the present results indicate that self bonded bioactive glass fiber scaffolds constructed of 13-93 and 45S5c with average porosity measurements between 44 and 63% and compressive strengths between 0.6 and 5.3 MPa have the ability to function *in-vivo*, promote bone tissue formation, possibly promote angiogenesis, and be biologically converted to calcium phosphate or resorbed by body fluids.

ACKNOWLEDGEMENTS

Support for this work was provided by a Missouri University of Science and Technology Graduate Research Assistantship. The authors would also like to thank Wesley Glick for assistance with the scaffold implantation.

REFERENCES

[1] L. Hench, Bioceramics, *J Am. Ceram. Soc.* **81** (1998) 1705.
[2] L. Gibson, M. Ashby, Cellular Solids – Structural properties 2nd edition. (Cambridge Solid State Science Series,1997) p. 429-452.
[3] J. Park, R. Lakes, Biomaterials: An Introduction 2nd edition. (Plenum Press, New York, NY, 1992) p. 197.
[4] L. Hench, J. Jones. Factors affecting the structure and properties of bioactive foam scaffolds for tissue engineering, *J Biomed Mater Res B: Appl Biomater* **68B** (2004) 36.
[5] S. Callcut, J. C. Knowles, Correlation between structure and compressive strength in a reticulate glass-reinforced hydroxyapatite foam, *J Mater Sci: Mater Med.* **13** (2002) 485.

6. H. Kim, J. Knowles, H. Kim, Hydroxyapatite porous scaffold engineered with biological polymer hybrid coating for antibiotic vancomycin release, *J Mater Sci: Mater Med.* **16** (2005) 189.

7. Q. Chen, I. Thompson, A. Boccaccini, 45S5 Bioglass®-derived glass–ceramic scaffolds for bone tissue engineering, *Biomaterials* **27** (2006) 2414.

8. E. Ebaretonbofa, J. Evans, High Porosity Hydroxyapatite Foam Scaffolds for Bone Substitutes, *J. Porous Mater.* **9** (2002) 257.

9. S. Ni, J. Chang, L. Chou, A Novel Bioactive Porous CaSiO₃ Scaffold for Bone Tissue Engineering, *J Biomed Mater Res A*; **76A** (2005) 196.

10. H. Ohgushi, M. Okumura, T. Yoshikawa, K. Inoue, N. Senpuku, S. Tamai, Bone Formation Process in porous Calcium Carbonate and Hydroxyapatite, *J. Biomed Mater Res.* **26** (1992) 885.

11. M. Rahaman, R. Brown, S. Bal, D. Day, Bioactive Glasses for Nonbearing Applications in Total Joint Replacement, *Seminars in Arthroplasty.* **17** (2006) 102.

12. R. Martin, M. Chapman, N. Sharkey, S. Zissimos, B. Bay, E. Shor, Bone Ingrowth and Mechanical Properties of Coralline Hydroxyapatite 1 Yr After Implantation, *Biomaterials* **14** (1993) 341.

13. T. Paatola, E. Pirhonen, P. Törmälä, Coating of bioactive glass (13-93) fibers with bioabsorbable polymer, *Bioceramics* **13** (2000) 717.

14. E. Pirhonen, L. Moimas, J. Haapanen, Porous Bioactive 3-D Glass Fiber Scaffolds for Tissue Engineering Applications Manufactured by Sintering Technique, *Key Engineering Mater.* **240-242** (2003) 237.

15. C. P.A.T. Klein, P. Patka, W. den Hollander, Macroporous Calcium Phosphate Bioceramics in Dog Femora: A Histological Study of Interface and Biodegration, *Biomaterials.* **10** (1989) 59.

16. R. Martin, M. Chapman, R. Holmes, D. Sartoris, E. Shors, J. Gordon, D. Heitter, N. Sharkey, A. Zissimos, Effects of Bone Ingrowth on the Strength and Non-Invasive Assessment of a Coralline Hydroxyapatite Material, *Biomaterials*, **10** (1989) 481.

17. M. Vogal, C. Voigt, U. Gross, C. Müller-Mai, In Vivo Comparison of Bioactive Glass Particles in Rabbits *Biomaterials* **22** (2001) 357.

18. A. Yao, D. Wang, W. Huang, Q. Fu, M. Rahaman, D. Day, In Vitro Bioactive Characteristics of Borate-Based Glasses with Controllable Degradation Behavior, *J. Am. Ceram. Soc.* **90** (2007) 303.

19. W. Huang, D. Day, K. Kittiratanapiboon, M. Rahaman, Kinetics and Mechanisms of the Conversion of Silicate (45S5), Borate, and Borosilicate Glasses to Hydroxyapatite in Dilute Phosphate Solution, *J. Mater. Sci: Mater Med.* **17** (2006) 583.

20. L. Hench, D. Clupper, Crystallization kinetics of tape cast bioactive glass 45S5, *J. Non-Crystalline Solids* **318** (2003) 43.

21. E. Soboleva, N. Yuritsyn, V. Ugolkov, Kinetics of Crystal Nucleation of Na₂O-2CaO-3SiO₂ Based Solid Solutions in Glasses of the Na₂SiO₃-CaSiO₃ Pseudobinary Join, *Glass Physics and Chem.* **6** (2004) 481.

22. V. Fokin, O. Potapov, E. Zanotto, F. Spiandorello, V. Ugolkov, B. Pevzner, Mutant Crystals in Na₂O-2CaO-3SiO₂ Glasses, *J Non-Crystalline Solids* **331** (2003) 240.

23. O. Peitl, E. Zanotto, L. Hench, Highly bioactive P₂O₅–Na₂O–CaO–SiO₂ glass–ceramics, *J Non-Crystalline Solids* **292** (2001) 115.

24. L. Hench, G. La Torre, O. Filho, E. Zanotto, United States Patent #5,981,412. (1999).

25. M. Brink, The Influence of Alkali and Alkaline Earths on the Working Range for Bioactive Glasses, *J. Biomed. Mater. Res.* **36** (1997) 109.

26. S. Jung, Silicate Based Bioactive Glass Fiber Scaffolds for Bone Tissue Regeneration, M.S. Thesis, Missouri University of Science and Technology (2007)

27. R. Brown, D. Day, T. Day, S. Jung, M. Rahaman, Q. Fu, Growth and Differentiation of Osteoblastic Cells on 13-93 Bioactive Glass Fiber Scaffolds, *Acta Biomater*, **4**, 387-96 (2008).

28. O. Laboux, N. Dion, V. Arana-Chavez, L-G Ste-Marie, A. Nanci, Microwave irradiation of ethanol-fixed bone improves preservation, reduces processing time, and allows both light and electron microscopy on the same sample, *J. Histochem. And Cytochem.* **52** (2004) 1267.

29. A. Leu, J. Leach, Proangiogenic Potential of Collagen/Bioactive Glass Substrate. *Pharma Res.*, 2008. **25**(5): p. 1222-1229.

30. J. Leach, Coating of VEGF-releasing scaffolds with bioactive glass for angiogenesis and bone regeneration. *Biomaterials*, 2006. **27**: p. 3249-3255.

31. K. Karlsson, Bioactivity of glass and bioactive glasses for bone repair, *Glass Technol.* **45** (2004) 157.

32. S. Radin, P. Ducheyne, S. Falaize, A. Hammond, *In vitro* transformation of bioactive glass granules into Ca-P shells, *J. Biomed. Mater. Res.* **49** (2000) 264.

ENZYME GRAFTING TO BIOACTIVE GLASSES

Enrica Vernè[*], Sara Ferraris[*], Chiara Vitale Brovarone[*], Silvia Spriano[*], Oana Bretcanu[*], Claudia Letizia Bianchi[§], Marco Morra[#], Clara Cassinelli[#]

*Material Science and Chemical Engineering Departement, Politecnico di Torino C.so Duca degli Abruzzi 24, 10129, Turin, Italy
§ Dept. Physical Chemistry and Electrochemistry, Milan University, V. Golgi 19, 20133 Milan (Italy)
#NobilBio Ricerche, V. Valcastellana 26, 14037 Portacomaro (AT)

ABSTRACT
The opportunity to obtain surfaces which are bioactive from both the inorganic and biological point of view represents an interesting solution in order to improve tissue integration of implants and bone regeneration. The aim of this research work is enzyme grafting onto bioactive glasses in order to promote also a biological bioactivity.

Two kinds of glasses, with different compositions and degree of bioactivity were employed as substrates. Samples have been firstly washed to expose hydroxyls groups and then silanized with 3-amminopropyltriethoxysilane in order to introduce amino groups. Finally, samples have been grafted with the enzyme alkaline phosphatase (ALP). Both silanized and only washed samples have been grafted with ALP. Different washings were performed after functionalization in order to study bonding stability. Hydroxyls exposition and silanization have been verified by means of contact angle measurements. ALP grafting has been studied by means of XPS analysis. After the addition of the specific substrate, the ALP activity was evaluated by UV-Vis spectroscopy. XPS spectra revealed ALP presence onto surfaces of grafted samples by enrichment in carbon and nitrogen and a decrease in the constituents' characteristic of the substrate. Peaks characteristic for the enzyme can be evidenced by the detailed study of carbon region. Enzymatic activity tests showed that ALP maintains its activity after grafting and also that enzyme activity is reduced but maintained after different washings.

INTRODUCTION

The most important goal of prosthetic surgery, both in orthopaedic and dental field, is the rehabilitation of injured bone with a functional and healthy one. In order to obtain this result a lot of solutions have been proposed and applied to improve tissue integration of implants.

Bioactive glasses and glass-ceramics are considered as optimal candidates as biomaterials for bone reconstruction because of their ability to form, in contact with physiological fluids, a hydroxyapatite or hydroxy-carbonate-apatite layer, analogue to the mineral phase of the bone able to guide new bone growth on material surface [1].

The ability to promote hydroxyapatite precipitation in contact with physiological solutions is generally called inorganic bioactivity. In addition to this bioactive behaviour the possibility to induce a specific biological response has been recently explored through the grafting of biomolecules (proteins, peptides, growth factors and enzymes) to material surface. Biomolecules are able to stimulate directly cells in order to promote new tissue regeneration. A lot of studies consider the grafting of extracellular matrix proteins (such as integrin) or peptides from them, to implant materials in order to mimic physiological adhesion of cells to the substrate [2]. On the other hand several research groups focused their attention on the opportunity to stimulate cells behaviour and improve tissue healing, through local delivery of growth factors [3].

In this research work the enzyme alkaline phosphatase (ALP) has been grafted to bioactive glasses surface in order to conjugate inorganic bioactivity characteristic of the substrate to a specific biological bioactivity introduced by the biomolecule.

ALP has been chosen because it can be a good model molecule since it is well known and effective techniques have been developed for its detection; at the same time it is involved in bone remodelling processes and it seems that it is able to improve bone mineralization.

MATERIALS AND METHODS

Two different bioactive glasses have been considered: SCNA and CEL2. Both glasses have been prepared by melting the precursors in a platinum crucible and quenching the melt on a brass plate obtaining glass bars, which have been subsequently annealed in furnace to relax residual stresses. Molar percentages of oxides are reported below.

SCNA: 57% SiO$_2$, 34% CaO, 6% NaO and 3% Al$_2$O$_3$

CEL2: 45% SiO$_2$, 3% P$_2$O$_5$, 26% CaO, 7% MgO, 15%Na$_2$O, 4K$_2$O

The first glass (SCNA) is characterized by a simple composition, a high stability and a low degree of bioactivity, while the second one (CEL2) has a more complex composition, is more reactive and presents a higher degree of bioactivity.

Glass bars have been cut with a mechanical cropper and polished with SiC abrasive papers in order to obtain homogeneous slices for the experiments.

The proposed surface modification includes three steps:
- surface exposition of active groups
- introduction of specific functional groups
- biomolecular anchoring

The first passage allows the exposition of hydroxyls groups on glass surface and includes a degreasing washing in acetone followed by three washings in double distilled water. All these washing have been performed in an ultrasonic bath [4].

The second step is optional and is related to the introduction of specific functional groups for molecular anchoring. In this work silanization with 3-aminopropyltriethoxysilane (APTES) has been considered in order to introduce amino groups on the glass surface [4,5]. Some silanized samples have been ultrasonic washed in TRIS-HCl solution in order to study silane-surface bonding stability.

Finally, both silanized and only hydroxylated samples have been grafted with ALP by soaking in ALP solution for 24 hours. ALP solution has been prepared dissolving alkaline phosphatase lyophilized powder in phosphate buffered saline (PBS) under magnetic stirring.

Functionalized samples have been washed in two different ways: gentle washing (sample dipping and removing from solution) and strong washing (2 min in ultrasonic bath). The first one allows simple removal of unbounded enzyme while the second one is performed in order to study bonding stability. Both washings have been performed in TRIS-HCl solution.

Static contact angle measurements with water droplet have been performed in order to verify both hydroxyls exposition and silanization. XPS analysis has been used for silane and enzyme detection. Finally enzymatic activity tests have been employed to determine ALP activity after grafting; functionalized samples have been introduced into a reactive mixture containing p-nitrophenilphosphate (a substrate for ALP); ALP has been quantified by determining the amount of the p-nitrophenol produced in the reaction. P-nitrophenol is yellow and can be quantified by UV-vis spectroscopy, measuring the absorbance at 405 nm.

Preliminary in vitro bioactivity tests on CEL2 samples in simulated body fluid (SBF) have been performed in order to study possible synergistic effect between inorganic and biological bioactivity.

RESULTS AND DISCUSSION

Washing procedure allows an effective hydroxyls exposition confirmed by a significant decrease in contact angles values for both glasses (Table 1). After silanization a considerable increase in contact angle could be observed for all samples (Table 1). The silane in fact introduces a hydrophobic layer onto material surface. Contact angle values of silanized samples do non change after TRIS-HCl ultrasonic washing for both glasses; this result confirms the stability of silane-surface bonding.

XPS analyses confirm the nitrogen enrichment of the samples surface after silanization, as shown for SCNA sample in Figure 1a as an example. Nitrogen amount is almost zero on samples after washings and it reaches a value comprised between 3 and 4 (atomic %) for silanized samples.

Table I: Contact angle after washings and silanization

	Contact Angle [°]					
	SCNA			CEL2		
	Glass	Acetone-water washings	Silanization	Glass	Acetone-water washings	Silanization
mean	69,76	26,38	75,82	52,76	22,71	68,94
stdev	4,82	2,86	3,50	6,69	0,79	0,17

XPS analysis after enzyme grafting underlines the increase of carbon and nitrogen, as shown for SCNA sample in Figure 1b as an example. The detailed analysis of carbon region shows a peak characteristic of hydrocarbons contaminants (284.6 eV) always present, but on ALP grafted samples (both silanized and not) two other peaks appear: one at 286 eV, typical of C-O and C-N bonds present into the enzyme and also a peak at 292-293 eV due to aromatic rings flattened on the surface attributable to phenylalanine, an amino-acid present in ALP [6].

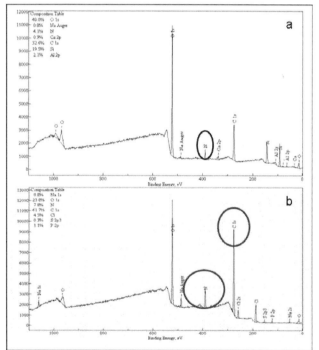

Figure 1: XPS survey spectra of a) silanized SCNA sample, b) silanized and ALP grafted SCNA sample

Enzymatic activity test reveal that ALP can be grafted to both glasses, despite of their reactivity, in an active state (Figure 2). Comparing absorbance values of functionalized samples with the ones of standard ALP solutions the enzymatic activity of samples is analogous to the one of solutions with an ALP concentration comprised between 0.5 and 1.0 mg/ml.

Looking at Figure 2 it can be noted that enzymatic activity is a bit lower on the more reactive glass (CEL2). This phenomenon can be explained considering that bioactivity reactions can interfere a little bit with bonding ones. In physiological solutions bioactive glasses react forming a silica gel layer which is then enriched in calcium and phosphorous and finally crystallizes in hydroxyapatite. So, on a bioactive surface at physiological pH (such as PBS one) hydroxyls can partially condense to form hydrated Si – O – Si species (silica gel) and hydroxyls condensation causes a lowering in the amount of active groups available for functionalization. Hydroxyls condensation is slower on the less reactive glass (SCNA) than on the more reactive one (CEL2), so a higher amount of hydroxyls is exposed and prone to silanization and biomolecule anchoring on SCNA than on CEL2. SEM observations of samples after functionalization confirm that silica gel has been formed on CEL2 samples but not on SCNA ones.

A strong ultrasonic washing significantly reduces the quantity of ALP on both glasses but an active part remains on the surface. The analysis of washing solutions reveals that the washed off of enzyme maintains its activity. It must be noted that ultrasonic stress is very different from physiological conditions, so a lower and more gradual release could be supposed in vivo. This data suggest that ALP from functionalized samples can stimulate tissue regeneration both from material surface and also in the implant site through the release in physiological fluids.

The difference in the amount of ALP that remains on silanized and only hydroxylated samples after ultrasonic washing is not significant so it seems that silanization does not significantly improve biomolecular grafting effectiveness.

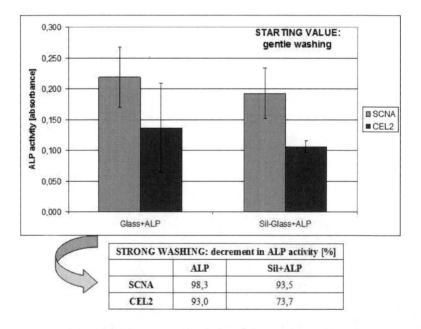

Figure 2: Enzymatic activity of functionalized glasses and decrement percentages after strong washing

Preliminary in vitro bioactivity tests suggest that ALP grafting onto bioactive glasses surface can enhance and quicken bioactivity kinetics.

CONCLUSION

It is possible to successfully graft the enzyme alkaline phosphatase onto different bioactive glasses despite of their reactivity. Anchoring can be performed both directly to hydroxyls groups, properly exposed on the surface, and through surface silanization; any significant difference among the two techniques can be detected in terms of grafting effectiveness and stability. ALP maintains its activity after grafting on all substrates.

A strong ultrasonic washing significantly reduce the amount of enzyme on the surface but an active part still remains anchored to the surface; the washed-off part maintains its activity in solution.

A synergistic effect is suggested between biological functionalization and inorganic bioactivity

Cellular tests are in progress in order to evaluate the effect of the grafted enzyme on the biological behaviour of materials.

REFERENCES

[1] W. Cao, L.L. Hench Bioactive materials. *Ceram Int*; **22**, 493-507 (1996)

[2] M. Morra, Biochemical modification of titanium surfaces: peptides and ECM proteins. *Eur Cell Mater*, **12**, 1-5 (2006)

[3] S-H Lee, H. Shin. Matrices and scaffold for delivery of bioactive molecules in bone and cartilage tissue engineering. *Adv Drug Deliv Rev*; **59**, 339-359 (2007)

[4] E. Vernè, C. Vitale-Brovarone, E. Bui, C.L. Bianchi, A.R. Boccaccini Surface functionalization of bioactive glasses. *J Biomed Mater Res*. DOI: 10.1002/jbm.a.32153.

[5] M. Ma, Y. Zang, W. Yu, H. Shen, H. Zhang, N. Ning, Preparation and characterization of magnetite nanoparticles coated by amino silane. *Colloid Surf A: Physiochem. Eng. Aspects*; **212**: 219-26 (2003)

[6] I.W. Nilsen, K. ØVerbø, R.L. Olsen. Thermolabile alkaline phosphatase from Northern shrimp (Pandalus borealis): protein cDNA sequence analyses. *Comp Biochem Phys*; **129**: 853-861 (2001)

FERRIMAGNETIC GLASS-CERAMICS FOR MAGNETIC INDUCTION HYPERTHERMIA

O. Bretcanu, S. Ferraris, M. Miola, E. Vernè
Materials Science and Chemical Engineering Department, Politecnico di Torino
c-so Duca degli Abruzzi 24, 10129, Turin, Italy
oana.bretcanu@polito.it

ABSTRACT
    This work is focused on the preparation and characterization of bioactive ferrimagnetic glass-ceramics containing antitumoral drugs for the treatment of cancer by magnetic induction hyperthermia. Under an alternating magnetic field, these biomaterials have a double action: they can generate heat and they can release the drug inside the tumour, enhancing the efficiency of the cancer treatment. These materials contain magnetite crystals, homogeneously distributed into a glass matrix. They are both bioactive and biocompatible to 3T3 murine fibroblast. The hysteresis loss and the specific power loss are compatible with the temperatures required for hyperthermic treatment of cancer. Two different drugs (doxorubicin and cisplatinum) were covalently bound to the glass-ceramic surface by a specific functionalisation process and they were released in an aqueous solution during two weeks. The uptake and release of doxorubicin and cisplatinum were evaluated on glass-ceramic powders, by using UV-visible spectrometry and graphite furnace atomic absorption spectroscopy.

INTRODUCTION
    Magnetic induction hyperthermia is one of the techniques used for cancer treatment. It consists in the destruction of the cancer cells by using magnetic materials which can generate heat under an alternating magnetic field. These magnetic materials are implanted into the tumour and heat due to their specific properties (usually hysteresis loss and eddy current loss) [1-3]. The heating temperature depends on the materials properties, magnetic field parameters (intensity, frequency) and tissue characteristics (cells type, blood flow, tissue density, thermal conductivity of the tissue etc).
    Hyperthermia can be applied together with chemotherapy, enhancing the drug activity directly *in situ* and limiting the gastrointestinal absorption of the drug.
    The aim of this work is development of a ferrimagnetic glass-ceramic for magnetic drug targeting by combining the hyperthermia with chemotherapy. Two antineoplastic drugs were selected: doxorubicin and cisplatinum. These drugs could be released directly on the desired tumour site by using the ferrimagnetic glass-ceramic microparticles as drug carrier.

MATERIALS AND METHODS
    Ferrimagnetic glass-ceramics belonging to the system SiO_2–Na_2O–CaO–P_2O_5–FeO–Fe_2O_3 were obtained by a traditional melting method. The composition was calculated in order to obtain a theoretical percent of magnetite of 45wt%.
    The glass-ceramic samples were characterised by X-ray diffraction (XRD), scanning electron microscopy (SEM) and differential thermal analysis (DTA). The magnetic hysteresis cycles were analysed using a vibrating sample magnetometer (VSM) with a maximum applied field of 12 kOe (956 kA/m), at room temperature, in quasi-static conditions. The specific power loss was determined from calorimetric measurements by using a magnetic induction furnace, with a magnetic field of 40kA/m and a frequency of 440 kHz. *In vitro* bioactivity was analysed by immersing the samples in a simulated body fluid (SBF) according to Kokubo protocol [4]. The samples removed from SBF after different times were analysed by SEM and XRD.
    Biocompatibility test was carried out by using 3T3 murine fibroblast cells. The cell adhesion was observed after 3h, while the cell proliferation was analysed after 72h. Both experiments (adhesion

and proliferation) were conducted in triplicate. The cell densities used for adhesion and proliferation were 20000 cells/cm^2 and 5000 cells/cm^2, respectively. All glass-ceramic samples were pre-conditioned 3 days in SBF before seeding. The cell number, cell morphology and cell vitality were analysed using a fluorescence microscope.

In order to bind the antitumoral drugs, the surface of the glass-ceramic materials was firstly activated by immersing the samples in distilled water and SBF, respectively. The drug uptake/release tests were made on functionalized glass-ceramic powders. The efficiency of the surface activation and drug release were investigated by contact angle measurements, UV-spectroscopy, spectrophotometry and X-ray photoemission spectroscopy (XPS).

RESULTS AND DISCUSSION

The XRD pattern of a glass-ceramic sample is shown in Figure 1. The spectrum puts in evidence the presence of a unique crystalline phase, magnetite, embedded in an amorphous residual phase. The glass transition temperature occurs at 650°C (DTA measurements).

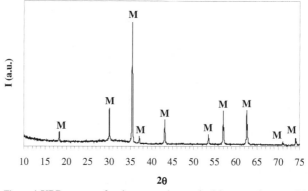

Figure 1 XRD spectra of a glass-ceramic sample (M=magnetite crystals)

SEM micrographs of a glass-ceramic sample after chemical etching are shown in Figure 2a and Figure 2b. The magnetite crystals are organised in small columns, homogeneously distributed into amorphous matrix. The crystals have an octahedral shape and a mean size of few microns. On the central part of Figure 2a it can be noticed a small column formed by octahedral crystals which are arranged one over the other. The mean crystals size is around 8µm. Figure 2b shows a characteristic magnetite crystal with octahedral shape; its average edge size is around 8µm. The EDS spectra of a crystal and a glass surface are shown in Figure 2c and Figure 2d, respectively. Chromium peaks come from the sample preparation process. The samples were coated by a chromium layer before observing them by SEM. The atomic ratio between iron and oxygen on the crystal region (Figure 2c) is similar to the one of magnetite. The glass region (Figure 2d) contains a small amount of iron oxide. Probably not all iron ions crystallised as magnetite; a small percentage of iron remains in the glass network. The EDS of the glass region identified all the elements present in the glass composition: sodium, calcium, silicon, phosphorus, iron and oxygen.

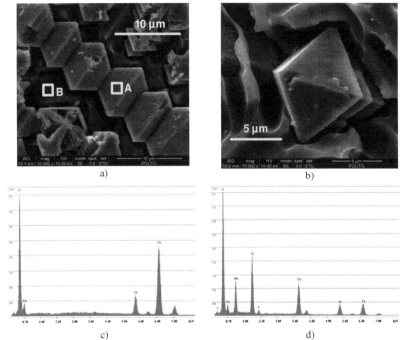

c)   d)

Figure 2a and 2b) SEM micrographs of a glass-ceramic sample c) EDS spectra of the A area marked on figure 2a (crystal) d) EDS spectra of the B area marked on figure 2a (glass).

Room temperature hysteresis loops of a glass-ceramic sample and pure magnetite under a magnetic field of 12 kOe are shown in Figure 3. The inset puts in evidence the central part of the hysteresis loops, emphasising the coercive field and the remanence magnetisation. The sample exhibits a magnetic behaviour characteristic of soft magnetic materials. It has a narrow coercivity and low remanence confronted with the commercial magnetite sample.

The specific power loss of these glass-ceramics is around 20W/g. The increase of temperature (ΔT) generated by 1g of these glass-ceramic samples under reported experimental conditions (40kA/m, 440kHz) is around 29°C. The hysteresis loss and the specific power loss are compatible with the temperatures required for hyperthermic treatment of neoplastic tissues.

These magnetic glass-ceramics are bioactive as hydroxylapatite crystals started to precipitate on their surface after 2 weeks of soaking in a simulated body fluid SBF. The bioactivity kinetics is rather slow confronted with high bioactive glasses (such as 45S5 Bioglass®).

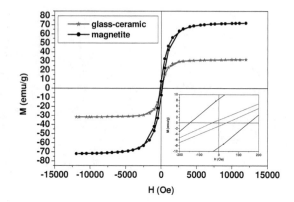

Figure 3 Room temperature hysteresis cycle of a glass-ceramic sample
and commercial magnetite powder

Biocompatibility tests show that the 3T3 murine fibroblast cells adhere and proliferate onto the glass-ceramic surface. Figure 4 presents the number of cells attached on the materials surface during the adhesion and proliferation tests. Polystyrene was used as control (CTR). The cells adhesion and proliferation on the glass-ceramic samples are lower confronted with the polystyrene control.

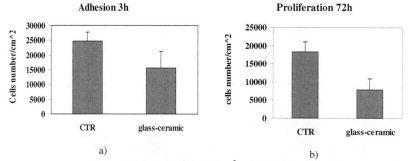

Figure 4 Number of 3T3 cells/cm^2 attached on the materials
a) after 3 hours (adhesion) b) after 72h (proliferation)

XPS and contact angle measurements showed that the samples were efficiently functionalised. The hydroxyl groups were activated on the glass-ceramic surface. The best pre-treatment method is dipping for 7days in distilled water at 37°C. The XPS spectra of the oxygen region before and after 7 days of pre-treatment in distilled water are shown in Figure 5. The deconvoluted oxygen peak presents three contributions: one characteristic of oxygen in different oxides, another one typical of oxygen in silica and the last one specific of oxygen in silanols, hydroxyl groups or silica gel. As can be seen, the peak corresponding to hydroxyl groups is higher on the sample obtained after the water pre-treatment. During the water dipping, the OH- groups are exposed and the surface is more reactive to the drug bonding.

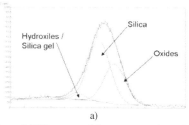 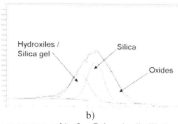

a)   b)

Figure 5 XPS spectra of the oxygen region a) before pre-treatment b) after 7 days in distilled water

The glass-ceramic powders previously pre-treated in distilled water for one week were dipped in the drug aqueous solutions for 1 day at 37°C. The drug released during 2 weeks of soaking in aqueous solutions is displayed in Figure 6. After 1day of soaking at 37°C, the glass-ceramic powders release around 15 wt% of doxorubicin and about 50 wt% of cisplatinum. The concentration of the doxorubicin released is very low, probably due to a strong covalent bond between the drug molecules and the hydroxyl groups exposed on the material surface.

The release kinetics is not uniform, showing irregular up and down peaks. The maximum amount of drugs released is obtained after 1day of dipping at 37°C. The drug release tests at 45°C are in progress.

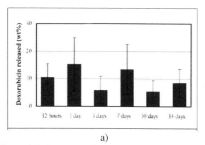 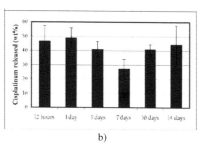

a)   b)

Figure 6 a) Amounts of doxorubicin (wt%) released during 2weeks b) amounts of cisplatinum (wt%) released during 2weeks

CONCLUSIONS

Ferrimagnetic glass-ceramics containing magnetite crystals were obtained by a traditional melting method. The specific power loss of these glass-ceramics is around 20W/g. These materials have a slow bioactive kinetics and are biocompatible to the 3T3 murine fibroblast cells. The hydroxyl groups were activated on the materials surface by soaking the glass-ceramic samples for one week into distilled water. After functionalisation, two different drugs were absorbed on the materials surface: doxorubicin and cisplatinum. Their release kinetics were studied by soaking the samples in aqueous solutions at 37°C for two weeks. The maximum amount of drugs released is obtained after 1day of dipping at 37°C. The release kinetics is irregular. The study of the drug release kinetics in function of the temperature is in progress.

ACKNOWLEDGEMENTS
The authors acknowledge P. Tiberto and P. Allia (National Institute of Metrology, Turin, Italy) for the magnetic measurements, Roberto Agu' and Giorgio Manfredi (Manfredi S.p.A, Pinerolo, Italy) for the calorimetric measurement facilities, Mario Cannas (Department of Medical Science, Human Anatomy, University of Eastern Piedmont, Novara, IT) for the in vitro tests, C. Bianchi and A. Naldoni (Department of Physical Chemistry and Electrochemistry, Milan University, Milan, Italy) for XPS measurements and G. Maina (Traumatology Orthopaedics and Occupational Medicine Department, Turin University, Turin, Italy) for the cisplatinum release test. This work was partially funded by the Piedmont Region project, "Functionalized ferrimagnetic materials for solid tumours", 2009.

REFERENCES
1. K. Ohura, T. Nakamura, T. Yamamuro, Y. Ebisawa, T. Kokubo, Y. Kotoura, et al. A bioactive ferromagnetic glass–ceramic for hyperthermia, *Ceramics in substitutive and reconstructive surgery*, Vincenzini, 131–136 (1991).
2. S.H. Oh, S.Y. Choi, Y.K. Lee, K.N. Kim, Research on annihilation of cancer cells by glass–ceramics for cancer treatment with external magnetic field. I. Preparation and citotoxicity. *Journal of Biomedical Materials Research* **54**, 360–365 (2001).
3. O. Bretcanu, E. Verné, M. Cöisson, P. Tiberto, P. Allia, Magnetic properties of the ferrimagnetic glass-ceramics for hyperthermia, *Journal of Magnetism and Magnetic Materials* **305**, 529-533 (2006)
4. T. Kokubo, H. Kushitani, S. Sakka, T. Kitsugi, T. Yamamuro, Solutions able to reproduce in vivo surface-structure changes in bioactive glass-ceramic A-W, *Journal of Biomedical Materials Research* **24**, 721-734 (1990).

RECENT RESEARCH ON COMPOSITION DEPENDENCE OF THE PROPERTIES OF BIOACTIVE GLASSES

Leena Hupa, Mikko Hupa
Process Chemistry Centre, Åbo Akademi University
Turku, Finland

ABSTRACT

In this paper we summarize past and present research at Åbo Akademi University on the measurement and modeling of the physical properties and the bioactive response of bioactive glasses as functions of the oxide composition. The first bioactive glass, 45S5 is still the fastest responding glass when implanted in the human body. However, 45S5 is apt to crystallize when heated, thus limiting the product types that can be made by hot-working. Thus, 45S5 is used mainly as monoliths or powdered glass. A lower bioactivity and a higher chemical durability may be valuable properties in clinical applications based on special product forms such as sintered porous implants or continuous fibers. The focus of the research being reviewed has been on developing tools for finding novel compositions of bioactive glasses with unique combinations of properties. The working properties were expressed with models for viscosity and crystallization characteristics of the glasses. Changes in simulated body fluid and in glasses after immersion were used to characterize and give basis to models of the *in vitro* bioactivity of the glasses as functions of the oxide composition. These models can be used for finding compositions of bioactive glasses with a desired product form (monoliths, fibers, microspheres or porous implants) and a desired *in vitro* bioactivity ranging from very bioactive to almost inert compositions.

BACKGROUND

Bioactive glasses have been a subject of intense research during the past forty years. The number of publications dealing with bioactive glasses has increased from less than 100 per year during the 1980's to more than 500 in 2008. Despite the extensive research only a few glass compositions are accepted for clinical use. The research of bioactive glasses at Åbo Akademi University started in the early 1980's as an attempt to develop phosphate opals, i.e. glasses which consisted of two phases, silica -rich matrix in which either glassy or crystalline calcium phosphate is suspended[1]. *In vivo* experiments of the compositions were done in cooperation with the Department of Dentistry at Turku University. The main goal was to find glass compositions for coatings, composites and glass-ceramics for hard tissue applications mainly in dentistry. Accordingly, the properties studied dealt with the use of glasses in dentistry. Thermal expansion was studied as the glass coating should fit to the metal substrate. Further, the thin glass coating on the metal should have a desired bioactivity but not be resorbed too rapidly. Thus, the dissolution behavior was also included in the studies. Parent glasses for glass-ceramics should have a controlled crystallization behavior to give a desired morphology. As the attempt was to sinter dense glass-ceramic monoliths from powdered glasses, also glass transition and crystallization temperatures were studied[1].

Simultaneously, research of the fundamental principles behind the material's ability to bond to bone was started at Åbo Akademi University. In general, the ability of a bioactive glass to bond to bone is described by the well-known reaction sequences within the glass and at its surface as suggested by Hench et al[2,3]. According to these, the bone-to-glass bonding is initiated by a rapid exchange of alkali ions with H^+ or H_3O^+ from solution followed by loss of soluble silica to the solution due to breaking of Si-O-Si bonds, and formation of Si-OH and $Si(OH)_4$. In the following step the silanol groups condense and repolymerize to a layer of silica gel on the glass surface. Formation of an amorphous calcium phosphate rich film on the top of the silica gel is supposed to depend on migration of Ca^{2+} and PO_4^{3-} ions through the silica layer[2,3]. The silica gel layer is supposed to grow through diffusion controlled alkali ion exchange, while the calcium phosphate layer grows by incorporation of soluble calcium phosphates from the solution. Finally, the amorphous calcium phosphate rich film crystallizes into hydroxyapatite by incorporation of OH^-, CO_3^{2-} or F^-. The small hydroxyapatite crystallites then interact with living cells and provide chemical bonding of the glass with bone. The initiation of calcium phosphate formation on the glass surface has been explained by a local alkaline pH at the glass-solution interface[4]. When bioactive glasses are implanted in bone, the reactions take place rapidly and a fairly strong bond with the bone forms within a few days. In general, bioactive glasses have a poor chemical resistance, which combined with the insufficient strength of the brittle material restrict the use of the glasses in load bearing applications.

Within the early bioactive glass research at Åbo Akademi University, models describing glass transition temperature and *in vivo* bone bonding in the system Na_2O-CaO-P_2O_5-B_2O_5-Al_2O_3-SiO_2 were developed by Andersson et al[1,5]. Alumina was added into the glasses in order to give a better chemical resistance, but was, however, found to prevent formation of the calcium phosphate layer, if the alumina content exceeded around 1 wt%. In the early 1990's one alumina-free composition from the Åbo Akademi glass series, S53P4, was successfully used in several *in vivo* studies dealing with non-load bearing applications[6-9]. Recently, some reports on the clinical use of S53P4 have been published[10-13]. Glass S53P4 is nowadays FDA approved and commercially available with the trade name BonAlive (Vivoxid Ltd).

The first bioactive glasses devitrified easily. In an attempt to prevent this, potassium and magnesium oxides were added, and alumina eliminated in the further research during 1990's. For this expanded system, a model describing the *in vivo* reactivity of the glasses as a function of the oxide composition was suggested by Brink et al[14]. From this series of experimental glasses, one composition denoted 13-93 has been employed in several *in vivo* studies[15-18]. This bioactive glass tolerates repeated hot-working with conventional forming methods. The glass can be drawn into continuous fibers, pressed and blown into various shapes and flame-sprayed into microspheres. Fibers of 13-93 have been studied e.g. in composites together with organic polymers[19,20].

Limited strength, one of the main problems encountered with the use of monoliths of bioactive glasses in load bearing applications, was a starting point for developing porous implants. The basic idea was to allow a three-dimensional formation of healing bone tissue by using implants with interconnected porosity. The implants were sintered of 13-93 microspheres[21]. *In vivo* studies indicated that the porous implant could be employed e.g. in reconstruction of deep osteochondral defects of weight-bearing joints[22]. Microroughening of the surface of porous implants by suitable chemical etching accelerated the early surface reactions *in vitro*[23]. These results were confirmed by studying the effect of microroughened microspheres of 13-93 in an intramedullary defect in rat tibia[24,25].

The bioactive glass research during 1980's and 1990's indicated that several physical properties as well as *in vitro* and *in vivo* properties could be expressed as functions of the oxide composition of the glasses. During the past years the need for tailoring glass compositions for various clinical applications has increased. Typical examples are glass compositions that can be sintered into scaffolds or drawn into reinforcing fibers to be used in composites together with organic polymers. Using glasses as fibers or thin-walled highly porous structures increases the surface area. Accordingly, the glasses are likely to react rapidly *in vitro* and *in vivo* and thus potentially lose their mechanical strength before new bone forms. This questions the use of bioactive glass fibers in reinforcing applications if not compositions which are non-toxic and adhere to bone but slowly resorb are used. So far, the clinical experiences of glasses only deal with non-load bearing applications. However, load bearing scaffolds and use of glasses also in soft tissue applications require that physical, chemical, *in vitro* as well as *in vivo* properties are properly understood.

NOVEL SERIES OF THIRTY GLASSES
There is an obvious need to tailor glass compositions which allow hot-working into various product forms and simultaneously show desired *in vivo* reactivity, either bioactivity or resorption. This has encouraged us to further develop our property models for the composition range that would be of interest in medical applications of glasses. One of the goals has been to calibrate the observed *in vitro* phenomena with *in vivo* responses of the same glasses. During the past years several models have been developed for calculating the physical and *in vitro* properties from the chemical composition[26-28]. These models were incorporated into an optimization routine for finding glasses which have desired hot-working properties and different levels of reactivity *in vitro*[29]. A comprehensive report of the *in vivo* calibration of nine of the new optimized compositions both in bone and soft tissue will be published shortly.

The different models describing glass properties from oxide composition were based on regression analysis of experimentally measured properties of thirty glass compositions within the system given below[26]. The minimum and maximum limits are given in wt%.

Na_2O	K_2O	MgO	CaO	B_2O_3	P_2O_5	SiO_2
5 – 25	0 – 15	0 – 6	15 – 25	0 – 4	0 – 4	50 - 65

The glass compositions were calculated statistically so that the correlation between the oxides was low to provide a sound basis for statistical evaluation of the measured data. Bioactive glasses 45S5, S53P4, 13-93 and 1-98 were used as references. The bioactivity of glass 1-98, a composition developed at Åbo Akademi University, has been confirmed *in vivo*[25]. Composition of 45S5 is outside the validity range of the models but serves as a reference for a glass that is highly bioactive and easily crystallizes during hot-working.

CRYSTALLIZATION
Bioactive glasses 45S5 and S53P4 were found to devitrify easily due to their low silica and high lime contents. Crystallization can be regarded either as a process that ruins glasses or as a process that can be utilized for manufacturing glass-ceramics of the parent glasses. Crystallization of bioactive glasses has several consequences: it hinders hot-working and forming of glasses to various products; it might be utilized to manufacture bioactive glass-ceramics with enhanced mechanical strength; it might change the bioactivity of the glass.

In Figure 1 is given a superimposition of our thirty experimental glasses and the four reference glasses in the Na_2O-CaO-SiO_2 phase diagram by Morey et al[30]. The compositions of the glasses were expressed as weight% of total alkalis, alkaline earths and silica. The figure was further simplified by omitting boron and phosphorus oxides. X-ray diffraction analysis of the heat treated experimental glasses indicated that the first crystalline phases in devitrification were either calcium silicate ($CaO \cdot SiO_2$) or sodium calcium silicate. In literature the compound $Na_2O \cdot 2CaO \cdot 3SiO_2$ has been suggested for crystals formed e.g. in 45S5. However, exact identification of the sodium calcium silicate crystals is rather difficult, because several strong peaks in the diffractograms are common to two or more of the phases[31]. Recently, solid solubility and continuous change in the composition of the crystal during the course of the crystallization for a glass which is just left of the stoichiometric $Na_2O \cdot 2CaO \cdot 3SiO_2$ ratio in the pseudo-binary system $CaO \cdot SiO_2$-$Na_2O \cdot SiO_2$ have been reported[32]. The experimental glasses in Figure 1 were given different symbols depending on the primary crystalline phases observed, squares for sodium calcium silicate and triangles for calcium silicate. Compositions close to the phase boundary between the two crystalline phases were marked by circles. The reference glasses were indicated by asterisks.

The reference glasses 45S5 and S53P4 are within the sodium calcium silicate field, while the glasses 13-93 and 1-98 are within the calcium silicate field. The simplified insertion of the experimental glasses into the ternary Na_2O-CaO-SiO_2 system suggests that their liquidus is within 1150-1350°C, i.e. at relatively high temperatures. The detected primary phases at devitrification, as indicated by the different symbols in Figure 1, seem nicely to follow the crystalline phases at liquidus. Thus, including potash and magnesia with soda and lime, respectively, seems to be a reasonable approximation for this type of presentation.

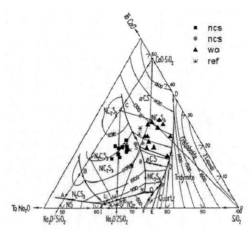

Figure 1. Observed primary phases in devitrification according to our own work[27] and theoretical primary phases formed at liquidus temperature[30]. ncs = $Na_2O\cdot2CaO\cdot3SiO_2$, wo = $CaO\cdot SiO_2$

We developed a model based on the primary phases of calcium silicate and sodium calcium silicate crystals. Formation of sodium calcium silicate was assigned the number 1 and the number 2 was assigned to calcium silicate. When calculating the primary crystalline phase in devitrification (PCT) using equation (1), PCT = 0.5-1.5 refers to formation of sodium calcium silicate and PCT = 1.5-2.5 suggests calcium silicate[27]. The oxides in the equation are expressed in wt%.

$$PCT = 0.018 - 0.107 \cdot x_{Na_2O} + 0.002 \cdot x^2{}_{Na_2O} + 0.052 \cdot x_{MgO} + 0.032 \cdot x^2{}_{P_2O_5} + 0.038 \cdot x_{SiO_2} \qquad (1)$$

Hydroxyapatite formation has been observed to be slower on partially crystallized glasses and glass-ceramics than on parent bioactive glasses[33,34]. We studied further the role of the primary crystal type on the *in vitro* bioactivity of the partially crystallized glasses[35]. Layer formation in simulated body fluid, SBF was slower on surfaces of samples that contained sodium calcium silicate crystals than on the parent glasses. On calcium silicate forming glasses layer formation was initiated by dissolution of the alkalis from the residual glassy phase. With prolonged immersion the calcium silicate crystals started to dissolve, thus delivering calcium ions for hydroxyapatite precipitation. Hydroxyapatite nucleated in the vicinity of the dissolving crystals and finally covered the whole surface. Besides calcium silicate or sodium calcium silicate, diopside formed in some magnesia containing glasses. These diopside crystals did not dissolve in SBF and clearly retarded the layer formation.

SINTERING PROPERTIES

The temperature window for liquid flow sintering of glasses is limited by glass transition and crystallization temperatures. Glass transition (T_g), onset of crystallization (T_x) and crystallization peak (T_p) temperatures were determined with thermal analysis[27,36-38]. The glasses could be divided into two groups; glasses with transition temperature around 500°C and onset of crystallization below 750°C, and glasses with transition temperature within 550 to 600°C and onset of crystallization around 900°C. All glasses within the first group crystallized into sodium calcium silicate, while glasses in the other group formed calcium silicate as the primary crystal. For the sodium calcium silicate glasses the sintering window is rather narrow, 150-200°C. The calcium silicate glasses allow sintering within a wider temperature range, 240-350°C. Sodium calcium silicate glasses showed bulk crystallization, while surface crystallization was observed for the calcium silicate glasses[38]. Equations (2) and (3) give the suggested models for glass transition temperature, T_g and crystallization temperature, T_x[27]. The oxides in the models are given in wt%.

$$T_g(°C) = -122.85 + 2.59 \cdot x_{K_2O} + 5.36 \cdot x_{MgO} + 9.37 \cdot x_{CaO} + 4.08 \cdot x_{B_2O_3} + 6.45 \cdot x_{P_2O_5} + 8.04 \cdot x_{SiO_2} \quad (2)$$

$$T_x(°C) = 191.70 - 9.51 \cdot x_{Na_2O} + 5.91 \cdot x_{MgO} + 5.93 \cdot x_{CaO} + 15.29 \cdot x_{B_2O_3} + 12.92 \cdot x_{P_2O_5} + 9.77 \cdot x_{SiO_2} \quad (3)$$

VISCOSITY

Viscosity controls all glass forming operations. Viscosity-temperature relationship for conventional soda-lime glasses is well-established. However, it is difficult to measure viscosity of bioactive glasses, as they easily crystallize above glass transition. Further, liquidus temperatures of bioactive glasses are at relatively low viscosities, thus limiting measurement of the high temperature viscosity values. Viscosity of 45S5 could be measured only down to around 1290°C and 0.76 Pa·s. Corresponding values for S53P4 were 1190°C and 5.2 Pa·s. Typically, the high temperature viscosities of the experimental glasses forming sodium calcium silicate crystals could be measured by rotation viscometer only at viscosities lower than 10 Pa·s, while the calcium silicate type glasses allowed measurement of higher viscosities to 10...1000 Pa·s, depending on glass composition[26]. However, as calcium silicate glasses show surface nucleation and slow crystal growth, for some compositions the highest viscosities with rotation viscometer might give values for a partially crystallized melt below liquidus. Combining low temperature viscosity values measured by hot stage microscopy and high temperature values by rotation viscometer enabled viscosity-temperature relationships for bioactive glasses[26]. Viscosity curves for 45S5, S53P4 and 13-93 are given in Figure 2a and the compositions of the glasses in Table I.

Table I. Oxide compositions of 45S5, S53P4 and 13-93 (wt%)

Glass	Na₂O	K₂O	MgO	CaO	P₂O₅	SiO₂
45S5	24.5			24.5	6	45
S53P4	23			20	4	53
13-93	6	12	5	20	4	53

Figure 2a shows that viscosity values of 45S5 and S53P4, both forming sodium calcium silicate crystals, could be measured only just above glass transition and at high temperatures. Within the intermediate temperatures the viscosity measurement suffered from crystallization. The calcium silicate type glass 13-93 allowed viscosity measurement almost within the whole temperature range. The measured values of the experimental glasses enabled modeling of several viscosity-temperature points as functions of oxide composition for the low and temperature ranges. The viscosity-temperature dependence is close to the Arrhenius type behavior as seen in Figure 2b.

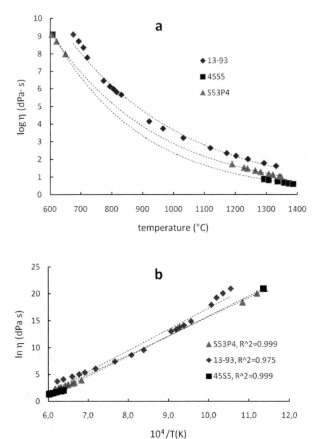

Figure 2. a) Experimental viscosity points for 45S5, S53P4 and 13-93 measured with hot stage microscopy at low temperatures and with rotation viscometer at high temperatures.
b) Arrhenius plot of the viscosity-temperature relationship for the glasses in Fig. 2a. R^2 is the correlation coefficient for the dashed straight lines.

Compositions forming calcium silicate are of interest for sintering of porous glassy implants, as their devitrification temperature is high enough to allow viscous flow before devitrification. Sintering of porous implants is possible when the viscosity is low enough (around 10^7 Pa·s) to allow softening and viscous flow of glass particles. A suitable temperature range for sintering can be approximated with equation (4) from the oxide composition of the glass:

$$T_{\eta=10^{8.35} dPa\cdot s}(°C) = 322.895 - 4.346 \cdot x_{Na_2O} + 2.643 \cdot x_{MgO} + 4.720 \cdot x_{CaO} + 4.202 \cdot x_{P_2O_5} + 5.201 \cdot x_{SiO_2} \qquad (4)$$

The maximum pore size of the implants increased with particle size, whereas the total porosity depended on sintering temperature and time[37]. Glasses within the calcium silicate field could also be drawn into continuous fibers both from performs and by down drawing from melt[39].

IN VITRO BIOACTIVITY

Several different ways were used to describe *in vitro* bioactivity of the experimental glasses. All these were based on experimental observations at immersion of glass plates in SBF at 37°C for 4 to 168 hours[28]. Changes both in the immersion solution and the glass surface were followed as functions of immersion time. A model for weight loss per surface area (Δm/A expressed as $\mu g/mm^2$) at 72 hours is given by equation (5)[28]. The oxides are given in wt%.

$$\ln(\frac{\Delta w}{A})_{72h} = -6.923 \cdot 10^{-1} + 8.921 \cdot 10^{-2} \cdot x_{Na_2O} + 6.915 \cdot 10^{-3} \cdot x_{K_2O}^2 + 1.810 \cdot 10^{-1} \cdot x_{MgO}$$
$$-1.967 \cdot 10^{-2} \cdot x_{MgO}^2 + 1.420 \cdot 10^{-3} \cdot x_{CaO}^2 + 3.364 \cdot 10^{-2} \cdot x_{B_2O_3}^2 + 1.962 \cdot 10^{-2} \cdot x_{P_2O_5}^2 \qquad (5)$$

Although the equation was achieved by statistical analysis of experimental data, it can be assumed to follow glass structure. The squared factor for potassium oxide compared to the linear dependence of sodium oxide suggests that replacing sodium with potassium decreases weight loss, as also can be expected from the sizes of the ions. Further, magnesia shows both linear and squared trends. This suggests that at higher contents magnesia acts as a network forming oxide, similarly to conventional soda-lime glasses. The squared factors for boron oxide and phosphorus pentoxide can be correlated with their action as network forming oxides in the bioactive glass structure.

Formation of hydroxyapatite layer *in vitro* is often taken as an indication of the bioactivity of glasses. The thickness and degree of surface coverage by the hydroxyapatite layer were used to develop a model, Bioactivity Number (BN) for the *in vitro* bioactivity as a function of the oxide composition, equation (6)[28]. Glasses with a uniform hydroxyapatite layer with a measured thickness > 0.9μm were given a relative bioactivity number BN=3. Glasses with incompletely developed hydroxyapatite layers with a layer thickness 0.1-0.9μm were described by BN=2. Glasses showing only spot like hydroxyapatite formation were denoted as BN=1. When using equation (6), BN > 2.5 indicates high bioactivity and BN < 1.5 low bioactivity. Intermediate bioactivity is suggested by values 1.5 < BN < 2.5. For the silica gel thickness on glasses at 72 h (TL$_{Si}$) best correlation was achieved when using relative values from 1 to 5, equation (7)[28]. The classifications were 1 for no silica layer at all, 2: 0.1-3.3μm, 3: 3.3-6.5μm, 4: 6.5-9.8μm and 5 for a silica layer thicker than 9.8μm. Further, the pH of SBF at different immersion times was expressed as functions of the oxide composition. Equation (8) gives the pH of SBF at 72 h immesion[28]. The oxides are given in wt%.

$$BN = 124.680 - 1.545 \cdot x_{Na_2O} - 4.500 \cdot 10^{-3} \cdot x_{Na_2O}^2 - 1.681 \cdot x_{K_2O} - 2.067 \cdot x_{MgO} + 4.199 \cdot 10^{-2} \cdot x_{MgO}^2$$
$$-1.661 \cdot x_{CaO} - 2.238 x_{B_2O_3} + 1.314 \cdot 10^{-1} \cdot x_{B_2O_3}^2 - 1.680 \cdot x_{P_2O_5} - 1.551 \cdot 10^{-2} \cdot x_{SiO_2}^2 \qquad (6)$$

$$TL_{Si} = 6.5387 + 7.847 \cdot 10^{-2} \cdot x_{Na_2O} + 8.725 \cdot 10^{-3} \cdot x_{K_2O}^2 - 5.747 \cdot 10^{-1} \cdot x_{B_2O_3} + 1.529 \cdot 10^{-1} \cdot x_{B_2O_3}^2$$
$$- 7.753 \cdot 10^{-1} \cdot x_{P_2O_5} + 2.207 \cdot 10^{-1} \cdot x_{P_2O_5}^2 - 1.573 \cdot 10^{-3} \cdot x_{SiO_2}^2 \tag{7}$$

$$pH_{72h} = 4.416 + 1.977 \cdot 10^{-3} \cdot x_{(Na_2O+K_2O)}^2 + 6.739 \cdot 10^{-2} \cdot x_{(MgO+CaO)} + 6.873 \cdot 10^{-2} \cdot x_{B_2O_3}$$
$$+ 5.908 \cdot 10^{-2} \cdot x_{P_2O_5} + 2.867 \cdot 10^{-4} \cdot x_{SiO_2}^2 \tag{8}$$

Table II gives a summarization of calculated *in vitro* values for glasses S53P4, 3-06 and 16-06. In the table is given also a calculated value of the *in vivo* bioactivity as expressed by the index of surface activity, ISA number. The ISA number is related with the formation of surface layers on glasses *in vivo*: inert glass =1, silica rich layer =2, layered structure = 3, bioactive =4, equation (9)[14]:

$$ISA = -8.4 + 79.0 \cdot \frac{x_{Na_2O} + x_{K_2O}}{x_{SiO_2}} - 1.8 \cdot \frac{(x_{Na_2O} + x_{K_2O})^2}{x_{SiO_2}} - 11.6 \cdot \frac{x_{MgO}}{x_{SiO_2}} - 9.2 \cdot \frac{x_{CaO}}{x_{SiO_2}} - 84.3 \cdot \frac{x_{B_2O_3}}{x_{SiO_2}}$$
$$+ 24.1 \cdot \frac{x_{B_2O_3}^2}{x_{SiO_2}} + 21.9 \cdot \frac{x_{P_2O_5}}{x_{SiO_2}} - 4.5 \cdot \frac{x_{P_2O_5}^2}{x_{SiO_2}} \tag{9}$$

Table II. Calculated values describing reactions of glasses S53P4, 3-06 and 16-06 caused by exposure in SBF[28] for 72h and after 8 weeks *in vivo*[14].

Glass	pH_{72}^{28}	$(\Delta m)/A^{28}$ ($\mu g \ mm^{-2}$)	TL_{Si}^{28}	BN^{28}	ISA^{14}
S53P4	7.85	9.4	4.4 medium level thickness	3.3 covering layer bioactive	4.7 bioactive
3-06*	8.02	10.4	3.5 medium level thickness	2.4 partial layer medium bioactivity	3.8 layered structure - bioactive
16-06**	7.42	3.5	1.3 thin layer	1.0 inert	0.8 inert

*24.6 Na$_2$O, 21.6 CaO, 1.3 B$_2$O$_3$, 2.5 P$_2$O$_5$, 50.0 wt% SiO$_2$
**5.0 Na$_2$O, 10.2 K$_2$O, 15.0 MgO, 2.0 B$_2$O$_3$, 4.0 P$_2$O$_5$, 63.8 wt%SiO$_2$

The same glasses have been tested in bone and soft tissue of rats[40]. The bone-bonding ability of the glasses was measured by push-out testing of truncated conical implants (height 7mm, average diameter 2.5mm) inserted into rat femur (cortical thickness around 1mm). In soft tissue testing the implants were inserted into the dorsal subcutaneous soft tissue or back muscles in rat. The soft tissue bonding was evaluated by pull-out testing. In both tissues the implantation time was 8 weeks. For the intramuscular site no measurable values were obtained. The maximum pull-out forces from bone and push-out forces from subcutaneous tissue are given in Table III.

Table III. Maximum push-out and pull-out forces of bone and soft-tissue implants after 8 weeks' implantation[40]. The values give mean and standard deviation of 8 implants (truncated cones, height 7mm, average diameter 2.5mm).

Glass	Site of implantation	
	Bone	Subcutaneous tissue
	Max. push-out force (N)	Max. pull-out force (N)
S53P4	49 ± 22	0.25 ± 0.19
3-06	68 ± 21	0.22 ± 0.18
16-06	4.0 ± 3	0.08 ± 0.11

The different calculated properties used to estimate the *in vitro* and *in vivo* reactivities of the glasses in Table II clearly correlate with both the bone and soft tissue observations. Glasses S53P4 and 3-06, which according to the calculated values would show high bioactivity also showed high push-out forces in the bone testing, i.e. were bioactive. Glass 16-06 was inert both according to the calculated values and the measured *in vivo* forces. The calculated values for the silica gel and hydroxyapatite layers as well as for the ISA number describing the *in vivo* bonding were higher for glass S53P4 than for 3-06. However, the push-out forces from bone suggested a somewhat stronger bone bonding of glass 3-06. The calculated weight loss per surface area and the increase in the pH of SBF were higher for 3-06 than for S53P4. These values would suggest that 3-04 dissolves somewhat faster than S53P4 in SBF. Whether this would also mean a faster bone bonding should be verified e.g. by analyzing the reaction layers formed on glass surfaces after *in vivo* testing.

The different models describing the *in vitro* bioactivity seem to nicely correlate with the *in vivo* observations. However, when developing glasses which should resorb slowly, e.g. reinforcing thin fibers in composites, also other factors besides formation of hydroxyapatite or silica gel layer should be taken into account. The models describing the relative numbers for the thickness of the silica gel layer and hydroxyapatite formation are valuable tools when tailoring glasses for applications where hydroxyapatite forms rapidly and thus gives bone bonding. For slower reacting glasses properties such as weight change or change in the ion composition of the solution appear more relevant. For slowly reacting glasses a longer test time both *in vitro* and *in vivo* should be used. This would require a development of the *in vitro* testing procedure to interfere with the long-term chemical durability. For novel applications of glasses, in e.g. load bearing scaffolds and reinforcing fibers in composites, the reactivity should be expressed using other criteria than formation of hydroxyapatite *in vitro*.

CONCLUSION

The non-crystalline structure of glasses offers a possibility of adjusting the physical and chemical properties additively within wide composition ranges by altering the oxide composition. The glassy nature also enables expressing several properties as functions of the oxide composition. The models developed recently for the composition range of bioactive glasses allow tailoring of glass compositions for a product form and reactivity suited for a specific clinical application. One of the main benefits with the approach developed is that most promising glass compositions for experimental testing can be effectively screened by computing using different optimization routines. The models describing the behavior of the melt, i.e. viscosity-temperature relationship and crystallization characteristics, are also valuable when adjusting the parameters at the glass manufacture.

Bioactive glasses are generally characterized by the ability to form a hydroxyapatite layer on the surface. The growth rate of this layer is supposed to correlate with the bone-bonding ability of the glass. The reactions leading to bone-bonding take place relatively rapidly after implantation *in vivo*. As the glass simultaneously dissolves, and ideally disappears totally with time, the strength of the implant decreases. For applications where the glass is designed to sustain some mechanical loading or act as a reinforcing component, compositions with higher silica content, and thus a lower ability to form hydroxyapatite layer are desired. An important issue for future research is to establish methods to measure and understand the long-term reactivity of glasses, both *in vitro* and *in vivo*. Ideally, glasses showing long-term chemical and mechanical durability are non-toxic, show good bone contact, are osteoconductive, and resorb slowly.

ACKNOWLEDGEMENTS

The authors want to thank Professors Kaj Karlsson and Antti Yli-Urpo and Doctors Örjan Andersson and Maria Brink for their pioneering bioactive glass research on which the present authors have been able to base their work.

Many people have contributed to the results presented in this paper. The excellent recent PhD Theses by Heimo Ylänen, Erik Vedel, Hanna Arstila and Zhang Di are sources for most of the results. Further, we want to thank our research partners, especially Professors Hannu Aro, Pertti Törmälä and Minna Kellomäki for the very fruitful collaboration within the many recent common projects.

This work is part of the activities of the Åbo Akademi Process Chemistry Centre within the Centre of Excellence program by the Academy of Finland. The research has further been supported by the Finnish Funding Agency of Technology and Innovation (Tekes), and the companies Vivoxid Ltd, Linvatec Biomaterials Ltd, Bioretrec Ltd and BbS-Bioactive Bone Substitutes.

REFERENCES

[1]Ö.H. Andersson, K.H. Karlsson, A. Yli-Urpo, K. Kangasniemi, Models for physical properties and bioactivity of phosphate opal glasses, *Glasstechn. Ber.*, **61**, 300-5 (1988).

[2]L.L. Hench, J. Wilson, Biocompatibility of silicates for medical use, Silicon Biochemistry, Ciba Foundation Symposium 121, John Wiley & Sons, 231-46 (1986).

[3]L.L. Hench, Ö.H. Andersson, Bioactive glasses, in An introduction to bioceramics, Eds. L.L. Hench and J. Wilson, World Scientific, 41-62 (1993).

[4]L.L. Hench, Bioactive Ceramics, *Ann. N.Y. Acad. Sci.*, **523**, 54-71 (1988).

[5]Ö.H. Andersson, G. Liu, K.H. Karlsson, L. Niemi, J. Miettinen, J. Juhanoja, In vivo behavior of glasses in the SiO_2-Na_2O-CaO-P_2O_5-Al_2O_3-B_2O_3 system. *J. Mater. Sci., Mater. Med.*, **1**, 219-27 (1990).

[6]J.T. Heikkilä, A.J. Aho, A. Yli-Urpo, Ö.H. Andersson, H.J. Aho, R.P. Happonen, Bioactive glass versus hydroxylapatite in reconstruction of osteochondral defects in the rabbit. *Acta Orthopaed. Scand.*, **64**, 678-82 (1993).

[7]J.T. Heikkilä, H.R. Salonen, A. Yli-Urpo, A.J. Aho, Long term behavior of bioactive glass cone and granules in rabbit bone. *Bioceramics*, **9**, 123-6 (1996).

[8]M.J. Peltola, J.T.K. Suonpää, Ö.H. Andersson, H.S. Määttänen, K.M.J. Aitasalo, A. Yli-Urpo, P.J. Laippala, In vitro model for frontal sinus obliteration with bioactive glass S53P4. *J. Biomed. Mater. Res.* **53**, 161-6 (2000).

[9]A.J. Aho, T. Tirri, J. Kukkonen, N. Strandberg, J. Rich, J. Seppälä, A. Yli-Urpo, Injectable bioactive glass/biodegradable polymer composite for bone and cartilage reconstruction: Concept and experimental outcome with thermoplastic composites of poly(.vepsiln.-caprolactone-co-D,L-lactide) and bioactive glass S53P4. *J. Mater. Sci., Mater. Med.*, **15**, 1165-73 (2004).

[10]N.C. Lindfors, J.T. Heikkilä, A.J. Aho, Long-term evaluation of blood silicon and osteocalcin in operatively treated patients with benign bone tumors using bioactive glass and autogenous bone. *J. Biomed. Mater. Res. Part B: Appl. Biomater.* **87B**, 73-6 (2008).

[11]N.C. Lindfors, J.T. Heikkilä, I. Koski, K. Mattila, A.J. Aho, Bioactive glass and autogenous bone as bone graft substitutes in benign bone tumors, *J. Biomed. Mater. Res. Part B: Appl. Biomater.*, in press.

[12]M. Peltola, K. Aitasalo, T. Tirri, J. Rekola, A. Puntala, Long-term tissue reactions of three biomaterials in craniofacial surgery, *Key Eng. Mater.*, **361-363**, 1343-6 (2008).

[13]A.J. Aho, E. Suominen, A. Alanen, A. Yli-Urpo, J. Knuuti, H.J. Aho, Remodeling of the tibia after grafting of a large cavity with particulate bioactive glass-hydroxylapatite--case report on treatment of fibrous dysplasia with 13 years' follow-up, *Acta Orthopaed. Scand.*, **74**, 766-70 (2003).

[14]M. Brink, T. Turunen, R.P. Happonen, A. Yli-Urpo, Compositional dependence of bioactivity of glasses in the system Na_2O-K_2O-MgO-CaO-B_2O_3-P_2O_5.SiO_2, *J. Biomed. Mater. Res.* **37**, 114-21 (1997).

[15]M. Brink, P. Laine, K. Narva, A. Yli-Urpo, Implantation of bioactive and inert glass fibers in rats - soft tissue response and short-term reactions of the glass. *Bioceramics*, **10**, 61-4 (1997).

[16]H. Niiranen, T. Pyhältö, P. Rokkanen, M. Kellomäki, P. Törmälä, In vitro and in vivo behavior of self-reinforced bioabsorbable polymer and self-reinforced bioabsorbable polymer/bioactive glass composites, *J. Biomed. Mater. Res., Part A*, **69A**, 699-708 (2004).

[17]A.J. Asikainen, J. Noponen, K. Mesimäki, O. Laitinen, J. Peltola, M. Pelto, M. Kellomäki, N. Ashammakhi, C. Lindqvist, R. Suuronen, Tyrosine derived polycarbonate membrane is useful for guided bone regeneration in rabbit mandibular defects, *J. Mater. Sci., Mater. Med.*16, 753-8 (2005).

[18]A.J. Asikainen, J. Hagström, T. Sorsa, J. Noponen, M. Kellomäki, H. Juuti, C. Lindqvist, J. Hietanen, R. Suuronen, Soft tissue reactions to bioactive glass 13-93 combined with chitosan. *J. Biomed. Mater. Res., Part A*, **83A**, 530-7 (2007).

[19]E. Pirhonen, H. Niiranen, T. Niemelä, M. Brink, P. Törmälä, Manufacturing, mechanical characterization, and in vitro performance of bioactive glass 13-93 fibers, *J. Biomed. Mater. Ress, Part B: Appl. Biomater.* **77B**, 227-33 (2006).

[20]T. Niemelä, H. Niiranen, M. Kellomäki, Self-reinforced composites of bioabsorbable polymer and bioactive glass with different bioactive glass contents. Part II: in vitro degradation. *Acta Biomater.* 4, 156-64 (2008).

[21]H. Ylänen, K.H. Karlsson, A. Itälä, H.T. Aro, Effect of immersion in SBF on porous bioactive bodies made by sintering bioactive glass microspheres. *J. Non-Cryst. Solids,* 275, 107-15 (2000).

[22]H.O. Ylänen, T. Helminen, A. Helminen, J. Rantakokko, K.H. Karlsson, H.T. Aro, Porous bioactive glass matrix in reconstruction of articular osteochondral defects, *Ann. Chir. Gynaecol.*, **88**, 237-45 (1999).

[23]A. Itälä, H.O. Ylänen, J. Yrjans, T. Heino, T. Hentunen, M. Hupa, H.T. Aro, Characterization of microrough bioactive glass surface: surface reactions and osteoblast responses in vitro. *J. Biomed. Mater. Res.* **62**, 404-11 (2002).

[24]A. Itälä, V.V. Välimäki, R. Kiviranta, H.O. Ylänen, M. Hupa, E. Vuorio, H.T. Aro, Molecular biologic comparison of new bone formation and resorption on microrough and smooth bioactive glass microspheres, *J. Biomed. Mater. Res., Part B: Appl. Biomater.*, **65B**, 163-70 (2003).

[25]A. Itälä, J. Koort, H.O. Ylänen, M. Hupa, H.T. Aro, Biologic significance of surface microroughing in bone incorporation of porous bioactive glass implants, *J. Biomed. Mater. Res. Part A,* **67**, 496-503 (2003).

[26]E. Vedel, H. Arstila, H. Ylänen, L. Hupa, M. Hupa, Predicting physical and chemical properties of bioactive glasses from chemical composition. Part I. Viscosity characteristics, *Glass Technol.: Eur. J. Glass Sci. Technol., A*, **49**, 251-9 (2008).

[27]H. Arstila, E. Vedel, L. Hupa, M. Hupa, Predicting physical and chemical properties of bioactive glasses from chemical composition. Part II. Devitrification characteristics, *Glass Technol.: Eur. J. Glass Sci. Technol., A*, **49**, 260-5 (2008).

[28]D. Zhang, E. Vedel, L. Hupa, H.T. Aro, Predicting physical and chemical properties of bioactive glasses from chemical composition. Part III. In vitro reactivity of glasses, *Glass Technol.: Eur. J. Glass Sci. Technol., A*, **50**, 1-8 (2009).

[29]E. Vedel, D. Zhang, H. Arstila, L. Hupa, M. Hupa, Predicting physical and chemical properties of bioactive glasses from chemical composition. Part IV. Tailoring compositions with desired properties, *Glass Technol.: Eur. J. Glass Sci. Technol., A*, **50**, 9-16 (2009).

[30]G.W. Morey, N.L. Bowen, Na_2O-CaO-SiO_2 system; in Levin, E.M. et al. *Phase diagrams for ceramists*, The American Ceramic Society, Ohio, 1964.

[31]R. Backman, K.H. Karlsson, M. Cable, N. Pennigton, Model for liquidus temperature of multi-component silicate glasses, *Phys. Chem. Glasses*, **38**, 103-9 (1997).

[32]V.M. Fokin, E.D. Zanotto, Continuous compositional changes of crystal and liquid during crystallization of a sodium calcium silicate glass, *J.Non-Cryst. Solids*, **353** 2459-68 (2007).

[33]O. Peitl, G.P. LaTorre, L.L. Hench, Effect of crystallisation on apatite-layer formation of bioactive glass 45S5, *J. Bio. Mater. Res.*, **30**, 509-14 (1996).

[34]O. Peitl, E.D. Zanotto, L.L. Hench, Highly bioactive P_2O_5–Na_2O–CaO–SiO_2 glass-ceramics. *J. Non-Cryst. Solids*, **292**, 115-26 (2001).

[35]H. Arstila, L. Hupa, K.H. Karlsson, M. Hupa, In vitro bioactivity of partially crystallized glasses, *Glass Technol.: Eur. J. Glass Sci. Technol.*, A, **48**, 196-9 (2007).

[36]H. Arstila, L. Hupa, E. Vedel, H. Ylänen, M. Hupa, The sintering range of porous bioactive glasses, *Glass Technol.*, **46**, 138-41 (2005).

[37]H. Arstila, E. Vedel, L. Hupa, Factors affecting crystallization of bioactive glasses, *J. Eur. Ceram. Soc.*, **27**, 1543-46 (2007).

[38]H. Arstila, L. Hupa, K.H. Karlsson, Influence of heat treatment on crystallization of bioactive glasses, *J. Non-Cryst. Solids*, **354**, 722-8 (2008).

[39]H. Arstila, M. Tukiainen, L. Hupa, H. Ylänen, M. Kellomäki, M. Hupa, In vitro reactivity of bioactive glass fibers. *Adv. Sci. Techn.* **49**, 246-51 (2006).

[40]D. Zhao, N. Moritz, E. Vedel, L. Hupa, H.T. Aro, Mechanical verification of soft-tissue attachment on bioactive glasses and titanium implants, *Acta Biomater.*, **4**, 1118-22 (2008).

Calcium Phosphates

ADSORPTION ON APATITIC CALCIUM PHOSPHATES: APPLICATIONS TO DRUG DELIVERY

F. Errassifi[1,2], A. Menbaoui[1,2], H. Autefage[1], L. Benaziz[1,2], S. Ouizat[1,2], V. Santran[1], S. Sarda[1], A. Lebugle[1], C. Combes[1], A. Barroug[2], H. Sfihi[3], C. Rey[1]

[1]University of Toulouse, CIRIMAT, UPS, INPT, CNRS, ENSIACET, Toulouse, France
[2]Departement de Chimie, Faculté des Sciences Semlalia,
[3] LPEM, ESPCI, and CSPBAT-CNRS, Université Paris Nord, Paris, France

ABSTRACT

The interfacial properties of apatitic calcium phosphate play a crucial role in calcified tissues and biomaterials. Generally the adsorption of molecules on apatitic calcium phosphates is considered to obey electrostatic interactions. The adsorption of various active molecules such as albumin, heparin, bisphosphonate, and growth factors like BMP-2, VEGF and FGF-2 on such surfaces seems to follow a similar pathway. In all cases the adsorption process can be well-described using Langmuir isotherms, although the adsorption process generally appears to be irreversible. Further studies of the adsorption reaction reveal, in most cases, an ion exchange mechanism involving the replacement of mineral ions on the apatite surface by molecular ions from the solution. In biomimetic apatite nanocrystals, considerable variations in the adsorption parameters are observed depending on the maturation time of the nanocrystals. As the maturation time increases, the adsorption affinity constant increases and the maximum amount adsorbed decreases. Knowledge of the adsorption process provides fundamental tools for the development of drug delivery systems using apatite materials. Applications based on the release of active molecules are examined in the case of apatite coatings.

INTRODUCTION

The adsorption of proteins and drugs on apatites are considered to play a determining role in mineralized tissues and substitutive biomaterials based on calcium phosphate apatites[1-3]. Several specific bone proteins such as osteocalcin, osteopontin, as well as generic proteins and biomolecules like albumin, phospholipids and others have been shown to adsorb on and interact strongly with apatite[4-9]. These phenomena are considered to control apatite crystal nucleation, growth and dissolution in vivo[5-8]. Many macromolecules have been classified as initiators or inhibitors of crystal growth although for a given substance the role is not as clear and could depend, for example, on the local concentration or on the state of the solute or solid fractions of that substance[10, 11]. In the biomaterial field it has been suggested that molecules that interact with calcium phosphates play a crucial role in the recruitment of cells and their expression and that they are strongly involved in the biointegration of implants[12, 13]. With regards the treatment of bone diseases, some bone-adsorbing molecules, like bisphosphonates, have been used to control bone turnover[9], and although their effect on bone cell behaviour has been widely documented, their biological activity has only recently been shown to be rather well correlated with their adsorption properties[14]. The surface properties of apatites are also widely used in the separation chromatography of proteins and, in fact, preliminary and development studies for these applications have prompted research on the adsorption on apatite surfaces[15]. The models proposed initially are still commonly used to explain the adsorption behaviour of apatites, essentially based on surface charge considerations[16, 17]. The aim of this report is to give several, non-exhaustive examples of different types of adsorption reactions on apatites and to present some possible applications especially for growth factor delivery. We will first describe adsorption experiments using small molecules, stressing the influence of functional groups on adsorption properties, and we will then analyse the effect of the characteristics of the adsorbent, especially the type of apatite used. Finally, we will describe the adsorption process of biological molecules and their applications.

159

MATERIALS AND METHODS

Adsorbents

The adsorption substrates were prepared according to published methods[18-21]. Briefly, almost stoichiometric apatite was prepared by double decomposition at boiling temperature and alkaline pH, between a calcium nitrate solution and an ammonium phosphate solution.

Apatite nanocrystals were also prepared by double decomposition but at room temperature and physiological pH[19].

After filtering and washing the samples were lyophilized. They were all characterized by X-ray diffraction, chemical analyses and FTIR spectroscopy. The specific surface area was determined by the BET method (nitrogen adsorption).

Adsorption experiments

Adsorption experiments were performed in batches at a constant temperature and known pH. A given amount of adsorbate was dissolved in aqueous solution at constant ionic strength (in KNO_3 solution). The apatite powder was added to the solution. After an equilibration period, determined from previous adsorption rate studies, the solutions were analysed. The adsorption isotherm was determined from the equilibrium concentration in solution. The calcium and phosphate content of the solutions was measured by atomic absorption spectroscopy and spectrophotometry of the phospho-vanado-molybdic acid respectively. Details are given elsewhere[19-21].

RESULTS

Adsorption of small molecules
- Serine and o-phosphoserine

The adsorption of these amino acids on apatite nanocrystals gives a good example of the effect of functional groups using simple molecules[21]. These amino acids are involved in bone matrix proteins and it has been shown that there is a relationship between the phosphorylation ratio of bone tissue proteins and the mineralization process. The adsorption isotherms of serine and phosphoserine on a nanocrystalline apatite are shown in Figure 1.

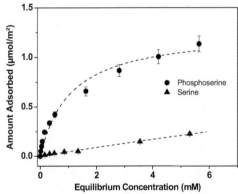

Figure 1: Example of the adsorption isotherm of serine and phosphoserine on nanocrystalline apatite (24 hours maturation time, Ca/P ratio: 1.42; Specific surface area 155 $m^2.g^{-1}$, pH 7.0, T=25 °C).

The adsorption of o-phosphoserine is well described by a classical Langmuir type isotherm and is characterized by a rather high affinity constant. In comparison, the adsorption of serine appeared to be much weaker and a plateau was not reached in the range of concentrations explored. The adsorption of serine can be described by a Freundlich type isotherm like that of other amino acids or simple molecules that weakly interact with apatite surfaces, such as glycine or acetic acid[22].

The very high affinity of phosphoserine for apatite surfaces compared to that of serine can be related to the existence of the phosphate group which can interact strongly with the apatite surface. Phosphoproteins, for example, have been shown to play a key role in apatite nucleation in bone tissues. In the present case, it has been shown by solid-state NMR that phosphorylated amino acids in particular seem to interact with nascent bone crystals in embryonic bone mineralization[23]. In the field of biomaterials, the attachment of methylmethacrylate monomers to calcium phosphate is considerably enhanced by the esterification of OH groups with phosphoric acid[24]. Among the other functional groups considered to interact strongly with apatite surfaces are silicate and sulphate. In general, carboxylate groups interact weakly except when they are associated in polyacids.

An interesting aspect of the adsorption of phosphoserine is that the maximum amount adsorbed can vary considerably depending on the pH or the presence of mineral ions in the solution. Thus in the presence of phosphate or carbonate ions the maximum amount adsorbed decreases whereas in the presence of calcium more molecules can be adsorbed at saturation.

The adsorption of phosphoserine cannot be reversed by simple dilution of the solution and the corresponding decrease in the equilibrium concentration of the adsorbate. This phenomenon is common to many adsorption processes on apatites.

- Adsorption of a bisphosphonate: risedronate
Bisphosphonates are among the most prescribed medications for several bone diseases and more specifically osteoporosis. The adsorption isotherm of a bisphosphonate, risedronate, on hydroxyapatite is shown in Figure 2.

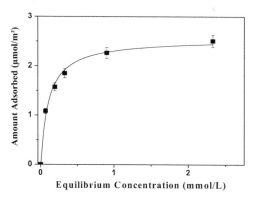

Figure 2: Adsorption isotherm of risedronate on well-crystallized hydroxyapatite
(Ca/P ratio 1.64, Specific surface area: 59 $m^2.g^{-1}$, pH=7.4, T=37 °C)

This also corresponds to Langmuir's isotherms. Analysis of the variation of the mineral ion content of the solution indicates that the adsorption of risedronate is associated with the release of phosphate ions (Figure 3) whereas the calcium concentration remains almost unchanged. These findings are analogous to those reported by others concerning the adsorption of another bisphosphonate, zoledronate, on calcium phosphates[25].

Given that the dissolution of hydroxyapatite in our system is congruent (i.e. the Ca/P ratio is the same for the solid and the solution), the calculations indicate that at most of the points on the isotherm there is one phosphate ion released for every risedronate molecule adsorbed. In this case the adsorption reaction corresponds in fact to an ion exchange:

$$[Ri^{3-}]_{sol} + [PO_4^{3-}]_{surf} \Leftrightarrow [Ri^{3-}]_{surf} + [PO_4^{3-}]_{sol}$$

in which the risedronate ions from the solution substitute for the phosphate ions on the apatite surface.

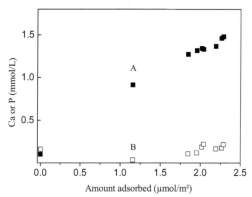

Figure 3: Variation in the calcium (B: white squares) and phosphate (A: black squares) content of the solution with the amount of risedronate adsorbed.

This reaction is responsible for the release of additional phosphate ions and the apparent non-congruent dissolution equilibrium of the apatite. According to this equilibrium, the calcium concentration should decrease as the phosphate in solution increases. This was not observed, and the stability and even the slight increase in the calcium content with the increase in risedronate in solution can be interpreted as an indication of the probable formation of ion pairs or complexes between risedronate molecules and calcium ions in solution.

An interesting aspect of risedronate adsorption is that the amount of molecules adsorbed at saturation can vary considerably depending on different parameters such as the temperature, pH, and initial phosphate and calcium content in the solution. Such variations cannot be explained by the chemical equilibrium proposed.

As previously noted the adsorption of risedronate is irreversible and the amount adsorbed is not altered by dilution of the adsorption solution or even by washing of the samples.

The adsorption of biological macromolecules is difficult to follow, as the possible interactions with a mineral surface are more numerous than those for simple molecules. One may regard a biological macromolecule as a charged object that can interact with a charged mineral surface depending respectively on the point of zero charge of the molecule and the surface charge of the mineral surface. Theoretical approaches, however, such as computational modeling, lead to multiple but precise bonding schemes involving identified groups of atoms of the molecule and of the mineral surface. In addition in many cases the adsorption of biological macromolecules is related to phosphate ion release[16], as for small molecules, and some of these molecules can be purified and separated by chromatography on apatite columns using phosphate solutions with a concentration gradient as for the mobile phase[26]. These observations suggest that the existence of specific chemical reactions with the apatite surface cannot be ignored in the description of adsorption processes.

Adsorption of biological macromolecules
-Bovine serum albumin (BSA)
Albumin is the most abundant protein in blood serum. It is also one of the major proteins in bone and it is thought to be involved in enamel formation. Several studies have examined the interaction of albumin with apatites[19, 27,28]. Like small molecules, BSA adsorbs on well-crystallized, almost stoichiometric apatite according to Langmuir's theory. However, the biological apatites are very different from stoichiometric ones: they are nanocrystalline and they exhibit a surface hydrated layer with mineral ions in non-apatitic environments. This layer is related to the precipitation process at physiological pH, and its structure, as well its extension, depend on the maturation time in solution[29]. The data obtained for BSA adsorption (and also for phosphoserine, and risedronate, data not shown) indicate an increase in the affinity constant and a decrease in the maximum amount adsorbed at saturation with the maturation time (Figure 4). These data illustrate the variability of adsorption on apatite substrates and the need to define and characterize precisely the type of apatite used and its surface.

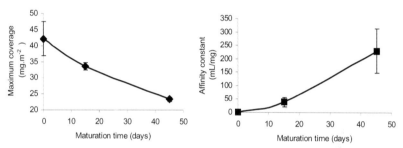

Figure 4: Variation in the adsorption characteristics of BSA on nanocrystalline apatites with maturation time[20].

-Heparin
Heparin is a calcium-complexing polysaccharide involved in the control of blood clot formation and has been proposed as a carrier of growth factors. The adsorption of heparin on well-crystallized stoichiometric apatite was investigated in detail. Like BSA and small molecules the adsorption equilibrium is reached relatively rapidly in about 20 minutes. The adsorption isotherms are Langmuirian (Figure 5).

Analyses of the mineral ions in solution showed complex variation (Figure 6). A constant increase in the concentration of phosphate ions as the amount of heparin in solution increases was observed. With regards the calcium concentration, a decrease occurred at low heparin levels, followed by an increase when the adsorption limit was reached. This behaviour differed from that of risedronate although the phosphate release seemed to increase with the amount adsorbed, and the U-shaped variation in the calcium content appeared rather strange. However, as for risedronate, the addition of calcium ions to the absorption solution increased the maximum amount adsorbed whereas the latter decreased in the presence of phosphate ions. Additional studies by solid state NMR on apatite nanocrystals showed that the adsorption of heparin was associated with a decrease in the surface HPO_4^{2-} ions.

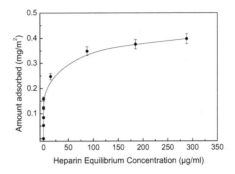

Figure 5: Adsorption isotherm of heparin on well-crystallized almost stoichiometric apatite. (Ca/P ratio 1.62, Specific surface area: 82 $m^2.g^{-1}$, pH=7.4, T=25 °C).

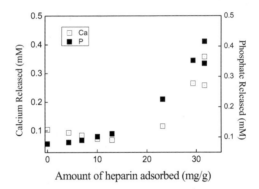

Figure 6: Variation in the calcium and phosphate concentrations in solution with heparin adsorption.

-rhBMP 2

BMP 2 is one of the most efficient growth factors that are involved in bone reconstruction. Only a few studies have been carried out on the adsorption of this molecule on apatites. Boix et al. produced a rather detailed study on the adsorption of BMP on well-crystallized apatites[30]. This molecule seems to follow the general characteristics of the adsorption of other molecules with a Langmuir type isotherm. As for many other molecules for which attachment to the apatite surface is determined by anionic carboxylate functions, the adsorption limit varies considerably with the mineral ions present in the solution. The increase in the calcium concentration in the adsorbate solution increases the adsorption limit whereas that of phosphate has the opposite effect. This growth factor can also be attached to nanocrystalline coatings[31]. The adsorption is also Langmurian, and most of the growth factor molecules are irreversibly bound and are not spontaneously released in cell culture media (Figure 7). This characteristic has also been observed for other growth factors like FGF 2 and VEGF[32,33].

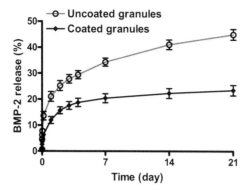

Figure 7: Release of rhBMP 2 in DMEM complete culture medium with 10% foetal calf serum from adsorbed rhBMP-2 on nanocrystalline coated and uncoated Biphasic Calcium Phosphate[31].

DISCUSSION
These few examples illustrate the variability of the adsorption processes on apatites. However, some common features can also be observed.

Adsorption isotherms
Generally the adsorption equilibrium of molecules that strongly interact with an apatite surface is well depicted by a Langmuir isotherm (C_e is the equilibrium concentration of the adsorbate, and Q is the amount adsorbed):

$$Q=NK_LC_e/(1+K_LC_e)$$

which is characterized by two adsorption parameters, the affinity constant (K_L) and the maximum coverage (N). The empirical Freundlich isotherm:

$$Q=K_FC_e^{1/n}$$

is generally found in the adsorption of molecules that are weakly bound to apatite surfaces, and are relatively far from the saturation limit. This was the case for serine in the example given here and also for other ions or molecules, including glycine and acetate ions.

In a recent study on adsorption Rill et al. tested different models for the adsorption of phosphonate-containing molecules on apatites[34]. The best fits were given by a Langmuir-Freundlich isotherm:

$$Q=N(K_{LF}C_e)^n/(1+(K_{LF}C_e)^n)$$

and a Tóth isotherm:

$$Q=[N(K_TC_e)^n/(1+(K_TC_e)^n)]^{1/n}$$

which take into account the existence of a distribution of adsorption energies on the apatite surface. However for the molecules of interest in this study, the simple Langmuir's isotherms are satisfactory. A major characteristic of adsorption on apatites is the frequent absence of reversibility, which is totally contradictory with the Langmuir theory of adsorption.

The adsorption process

Historically two types of adsorption sites have been recognized, on apatite calcium surface sites and phosphate surface sites respectively. Negatively charged molecules (or moieties) were considered to bind to calcium sites and positively charged ones to phosphate sites. The adsorption of negative molecules seems particularly important and is involved in most chromatographic uses of apatites and also in most studies involving molecules of biological interest. The zeta potential of apatites has been considered to play a major role in the adsorption process and most adsorption behaviours were, and still are, related to this surface parameter, although the correlation between this global physical parameter describing the surface charge of apatite particles and chemical interactions with specific ions at an atomic level is not straightforward. It appears that the global charge of a particle does not give any information on the local distribution of cations and anions on a mineral surface and in fact the zeta potential does not appear to be a decisive parameter: for example the adsorption of positive molecules on a positive apatite surface and inversely negative molecules on a negative apatite surface has been described[35, 36]. Similarly the isoelectric point of proteins, by itself, does not give specific information on the ionization state of the different functions of a molecule and should be considered with caution when predicting the adsorption behaviour related to a chemical interaction.

In several of the examples given in this report the adsorption of negative molecular ions has been related to the release of phosphate ions, demonstrating a chemical ion exchange reaction with the apatite surface. This process has been described in different instances, as for example in the adsorption of polyacids[37], citrate ions[38], zoledronate[25], or serum proteins on apatites[16]. In the case of risedronate, the adsorption of one molecule led to the removal of one phosphate ion from the apatite surface. This chemical process seems rather general and supports a chemical description of the adsorption reaction rather than a physical one based on global electrostatic interactions. Adsorption could thus be represented, for example, as a chemical equilibrium as follows (in the case of a one-to-one exchange):

$$A_{sol} + P_{ap} \Leftrightarrow A_{ap} + P_{sol} \quad \text{(reaction 1)}$$

with A being the adsorbate in a given ionic state either in solution (sol) or on the apatite surface (ap), and P being the phosphate group in a given ionic state either in solution or on the apatite surface. Other ion exchange possibilities have been discussed by Misra in the case of citrate adsorption[38]. These descriptions seem consistent with the use of apatites in chromatography where negatively charged molecules are desorbed using phosphate solutions (and positively charged ones using calcium solutions).

The peculiar behaviour of heparin deserves some additional explanation. This molecule can give soluble complexes with calcium ions. In the presence of apatite, calcium ions are complexed by heparin and apatite dissolves to reach a solubility equilibrium. During the initial rising part of the adsorption isotherm these complexes are bound to the apatite surface and phosphate ions are released. As surface saturation occurs, the addition of heparin still results in Ca-complex formation but the complex cannot attach to the saturated surface and remains in solution, resulting in an increase in both calcium and phosphate concentrations in the solution. This behaviour could also occur during the adsorption of citrate anions on apatites or more generally during the adsorption of calcium-complexing molecules.

In addition to complexation, other processes can disturb the adsorption equilibrium and should thus be taken into account: the possible precipitation of calcium (or phosphate) salts of the adsorbate, the dissolution equilibrium of the apatite, and the surface hydrolysis reactions of apatite.

Insoluble calcium salts may sometimes form and precipitate with the adsorbate molecules, depending on their respective concentrations. The apatite phase is then progressively destroyed as the new phase forms, leading to the release of phosphate into the solution until a new equilibrium is reached. This event does not generally give a Langmuir type isotherm with a typical saturation domain. On the contrary, the addition of aliquots of adsorbate solution results in precipitation of the calcium salt of the adsorbate and more destruction of the apatite. In some cases however this process could result in a Langmuir-like adsorption behaviour, with an apparent saturation domain if there is coating of the apatite surface by the new phase, thus preventing further reaction. Observation of the solid phase by X-ray diffraction or electron microscopy generally allows the formation of new phases to be detected.

The dissolution equilibrium of apatite also has to be considered. Apatite dissolution is generally congruent when the apatite/solution ratio is small enough and surface reactions can be neglected. However there are some conceptual divergences on the solubility product of hydroxyapatite that are most often related to the inhomogeneity of composition and non-stoichiometry. In the case of nanocrystalline apatites the dissolution behaviour is more complex than for stoichiometric apatites and it can probably be represented by a metastable equilibrium solubility (MES)[39] showing a dependency of the apparent equilibrium concentrations on the amount dissolved, although the dissolution still remains congruent. The amount of calcium and phosphate in the equilibrium solution depends also, in these cases, on the maturation stage of the nanocrystals. The result of dissolution equilibria is the presence of calcium and phosphate ions in the solution, which may interfere with the ion-exchange adsorption equilibrium.

In addition, in cases in which the amount of apatite with a high specific surface area is elevated in the solution, hydrolysis of the phosphate ions on the surface may occur, associated with a release of calcium into the solution. This phenomenon leads to an apparent non-congruent dissolution equilibrium, which appears to be mainly related to a surface equilibration reaction and thus depends on the solid/solution ratio. As adsorption studies are often made with rather large solid/solution ratios, this phenomenon leads to calcium and phosphate ions in solution in a ratio that differs from that in the solid, which may also interfere with the adsorption parameters. To avoid these anomalies, pre-equilibration of the apatite with a solution at the same pH and ionic strength as the adsorption solution can be performed, before placing the apatite in contact with the adsorption solution.

The description of adsorption as an ion exchange reaction enables the explanation of several characteristics of the adsorption of charged molecules on apatites and especially the apparent irreversibility of the adsorption process.

Irreversibility of the adsorption process

Langmuir adsorption is generally an equilibrium and thus is a reversible process. The most convenient way to determine the reversibility of adsorption is to dilute the solution in equilibrium with the apatite surface. Then the concentration of the adsorbate is reduced and a new equilibrium develops, resulting in the release of molecules from the solid surface. The total or partial irreversibility of adsorption on apatites upon solution dilution is a common observation that has received several explanations such as a kinetic limitation due to the multiple bonds of macromolecules, necessitating the rupture of all of them to release the molecule, a change in configuration and surface properties. These possibilities may of course exist, but irreversibility also occurs with simple molecules such as phosphoserine or bisphosphonate, for which none of these explanations seems satisfactory. With regards the ion exchange equilibrium reaction, it is apparent that the adsorption process is not only related to the concentration of the molecule in solution but also to that of the mineral ion that is displaced. From reaction 1 we can write:

$$K_{ex}= P_{sol}.Q/(C_e(N-Q))$$

where K_{ex} is the equilibrium constant of the exchange reaction, Q is the amount adsorbed, C_e is the equilibrium adsorbate concentration, P_{sol} is the phosphate concentration in solution, and N is the total number of adsorption sites (i.e. the maximum number of molecules adsorbed). This equation can be transformed into a Langmuir type isotherm equation:

$$Q= NK_{ex}(C_e/P_{sol})/(1+K_{ex}(C_e/P_{sol}))$$

The equilibrium concentration of the adsorbate (C_e) in the traditional Langmuir equilibrium is simply replaced by the ratio C_e/P_{sol} in the exchange reaction equilibrium. This model fits well with the data points obtained in our studies, and in addition it simply explains the apparent irreversibility on diluting the equilibrated solution. As dilution involves a reduction of both concentrations (C_e and P_{sol}) in equal proportion, the conditions of the equilibrium are roughly unchanged (if the very weak dissolution of apatite to re-establish the solubility equilibrium and hence the corresponding slight increase of calcium and phosphate in solution are ignored). The recently reported observation that the reversibility of the adsorption of phosphonate complexes[34] on apatite following dilution was dependent on the equilibrium concentration of the adsorbate in solution can be explained by the same arguments. In these experiments the authors tested the reversibility at two points of the isotherm corresponding to low and high amounts of adsorption. Dilution was performed with a solution saturated with respect to the apatite to avoid any dissolution of the adsorbent. Assuming an exchange reaction of a complex molecule with a phosphate ion, dilution with a saturated solution caused a relative increase in phosphate ions in solution that was more important in the solution equilibrated with a low amount adsorbed, and thus with the lowest amount of exchanged phosphate, than for the solution at a higher equilibrium concentration and a higher amount of phosphate in solution. Thus the reverse reaction appeared slightly stronger, in relative terms, at the low adsorbate equilibrium concentration. Similarly, washing an apatite sample with adsorbed molecules resulting from an ion exchange equilibrium does not permit any reverse ion exchange and no desorption should be expected, except for slight release effects due to the weak dissolution of the apatite.

An equivalent, though slightly different, description of the exchange reaction can be proposed by taking into account the solubility equilibrium of the apatite:

$$(Ca)^R(P)=K_{sp,pp}$$

where $K_{sp,pp}$ is the solubility product of the apatite corresponding to one phosphate and R is the Ca/P ratio of the apatite, assuming congruent dissolution (the pH is considered constant, for example at physiological pH, and the OH^- contribution to the solubility is therefore also constant and included in the solubility product).

Thus the adsorption isotherm may also be written as:

$$Q= NK_{ex}(C_e(Ca)^R/K_{sp})/(1+K_{ex}(C_e(Ca)^R/K_{sp}))$$

or:

$$Q= NK_b (C_e.(Ca)^R/(1+K_bC_e.(Ca)^R)$$

where $K_b=K_{ex}/K_{sp,pp}$. The adsorption process can thus be described as the mobilization of calcium ions from the solution to serve as the binding agent of the molecule with the mineral, instead of for phosphate exchange. The corresponding chemical equation can be written as:

$$A_{sol} + RCa_{sol} + S_{ap} \Leftrightarrow A\text{-}RCa\text{-}S_{ap}$$

where S_{ap} represents a binding site on the apatite. These raw equations have to be adapted to each case, and there can be, for example, one binding calcium ion per molecule, which could be described in the equivalent ion exchange process as the replacement of a fraction of phosphate ion. These descriptions (ion exchange with phosphate or calcium binding) are totally equivalent from a chemical point of view although they have to be adapted to take into account the number of ions involved and their charges.

The adsorption process does not differ significantly between well-crystallized apatites and nanocrystalline apatites. An interesting aspect is the variation of the adsorption parameters with the characteristics of the nanocrystals.

Adsorption on apatite nanocrystals and ageing
An interesting problem in nanocrystalline apatites is the possible change of surface properties with the maturation state of the nanocrystals.

The available data on adsorption properties of apatite nanocrystals are very scarce, but those obtained with several molecules such as bovine serum albumin, phosphoserine, and risedronate suggest a general behaviour, and such findings (reported in Figure 4, in the case of albumin) have also been observed in other cases: the number of adsorption sites decreases when the maturation time of the nanocrystals increases, and the affinity constant increases.

Variation in the affinity has been related to a decrease in the surface energy as the hydrated layer becomes progressively transformed into apatite. Although it is very difficult to determine the surface energy of nanocrystals, several measurements on Ca-P materials have shown that hydrated Ca-P salts always have a much lower surface energy than anhydrous ones[40]. Thus, the hydrated layer, which most probably reduces the surface energy of the nanocrystals in aqueous media, also weakens, to some extent, the interfacial energy released by protein adsorption. Another explanation could be that ions in the hydrated layer become less exchangeable as the maturation time increases, as effectively observed[29].

The decrease in the maximum number of adsorbed molecules with maturation is more difficult to explain. It does not seem to be related to a decrease in the specific surface area of the crystals and the rate of coverage does not support the formation of multilayers of proteins on immature crystals. A different conformation could be suggested in the case of albumin. However even very small molecules such as phosphoserine, for which conformation changes cannot be proposed, exhibits the same behaviour. The adsorption process by ion displacement suggests that the amount of protein incorporation could be related to the amount of labile mineral ions available in the hydrated layer and thus to the extension of the hydrated layer. The adsorption process could thus involve the whole hydrated layer and possibly its re-structuring and stabilization. As a support for this hypothesis it can be noted that the adsorption limit is highly correlated with the development of the hydrated layer on nanocrystals, as shown for example for the adsorption of phosphoserine (Figure 8). It has also been shown that the development of the hydrated layer is related to the ion exchange capability, and a parallel seems to exist between ion exchange and adsorption.

Another common feature of adsorption on apatites is the variability of the maximum adsorbed amount with several physical chemical parameters.

Maximum amount adsorbed
The maximum amount adsorbed is supposed to correspond to the saturation of the surface

adsorption sites. Questions regarding the variations in this quantity are generally answered by reference to the surface charge model and its alterations. For example an increase in the pH produces a decrease in the surface charge and in the adsorption capabilities of negatively charged molecules. Similarly an increase in the calcium ion concentration in solution increases the surface charge and favours the adsorption of negatively charged molecules whereas an increase in the phosphate concentration has the reverse effect. As already mentioned, this global model does not consider the fact that the most important adsorption processes on apatite involve strong local interactions with individual ions on the surface and not simply a global electrostatic interaction between a charged surface and a charged molecule.

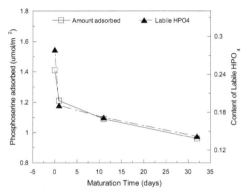

Figure 8: Correlation between the amount of phosphoserine adsorbed and the development of the hydrated layer as measured by the intensity of the surface HPO_4^{2-} determined by FTIR spectroscopy.

In terms of the proposed ion exchange reactions, the chemical equilibria involved cannot explain the variation in the adsorption limit: when the concentration of absorbate increases to large values, all sites should be occupied. Such variations suppose that the maximum number of sites that can be occupied (N) varies according to different factors. It is obvious that for most molecules, and this is especially true for macromolecules, the size is much larger than that of the phosphate or calcium ion exchanged by specific molecular functions, and the total replacement of all surface ions by these molecular functions is not possible. Thus there are always more possibilities for exchange and attachments than are feasible due to steric constraints. The effect of phosphate and calcium ions in the adsorption solution is always similar with respect to the adsorption limit of negatively charged molecules: phosphate ions decrease the maximum amount of adsorbed molecules compared to a solution containing only the adsorbate, and calcium ions increase it. These added mineral ions affect the adsorption limit within a wide range of values (from 1 to 6 for example in the case of heparin). If we compare this with the variations observed for nanocrystals at different maturation stages it can be suggested that the mineral ions in solution may affect the hydrated layer at the surface of the apatite crystals. The strength of attachment and surface mobility of the adsorbed molecules also have to be considered and could be involved in the reorganization of molecules on the surface.

Consequences for bone mineral

Two main roles have been assigned to the interactions of proteins with apatite crystals in biological systems: the initiation of crystal nucleation and inhibition of crystal growth[41]. Several in vitro experiments have provided support for this notion. However the classification of proteins as initiators or inhibitors of mineral formation suffers from numerous exceptions. It has been shown that the role of proteins may depend on their state and configuration. Thus, the same protein may act as an inhibitor when it is in solution and as an initiator when it is bound to a substrate. In addition most adsorbing proteins seem to show both aspects depending on their concentration[10]: at low concentrations they accelerate crystal formation whereas at high concentrations they act as inhibitors.

Most mineral binding proteins are rich in anionic groups. Phosphoproteins in particular have been suggested to bind to calcium and to initiate the first nuclei for mineral formation[42]. Due to their lower mobility compared to that of mineral ions and the relatively fast rate of formation of nanocrystals, proteins can only play a role when they are already in situ. In addition to their role in the nucleation and growth of mineral crystals, proteins may also be used to control mineral maturation. Maturation of the nanocrystals considerably reduces their surface reactivity and their action in homeostasis. Remodeling appears, therefore, to be necessary for the organism to maintain the ion reservoir function of bone. The wide variation in the surface properties of the nanocrystals with maturation and especially the change in adsorption properties suggest that some proteins could act as signals between mineral crystals and cells to control the maturation state and possibly initiate the remodeling process. However, this hypothesis requires verification. The variation in the adsorption properties of bone nanocrystals with time is rarely considered in a dynamic approach to the system; but, given the very large specific surface area of bone crystals, it could be fundamental.

The interactions between nanocrystals and the collagen matrix have not yet been elucidated at the atomic level. It seems probable that ion mobility in the hydrated layer could facilitate intimate contact between the two main constituents of bone and could influence the mechanical properties of the tissue. The fact that the mechanical properties of bone depend strongly on its hydration state supports this hypothesis.

With regards the specific mode of interaction of drugs like bisphophonates with apatites, the study of the adsorption process allows some findings to be explained. The absence of accumulation of the drug in bone even after years of use, for example, is related to the adsorption process and the exchange equilibrium: once equilibrium concentrations have been reached in bone tissue the loading cannot continue. This would not be the case if a dissolution re-precipitation process were involved. Release after the interruption of treatment depends on the phosphate concentration in bone fluids; as it is very low this release is very slow.

It seems important to consider the dynamics of bone tissue, not only in terms of bone cell behaviour but also as a regulating system involving interfacial interactions with the mineral. Without taking this aspect into account the approach will necessarily be incomplete.

Consequences for biomaterials

Adsorption on apatites and other calcium phosphates may find several uses in biomaterials[1]. In general, active molecules such as growth factors are associated with biomaterials by impregnation, often followed by drying. However this easy, rapid and inexpensive method does not allow the precise control of the physical state of the active molecules, which may appear partly crystallized, partly adsorbed or partly precipitated as calcium or phosphate salts or more likely a combination of these possible states depending on the conditions of preparation and drying. Thus the release rate is difficult to predict; it can be rapid and only determined by the diffusion properties of the molecule if no interactions with the biomaterial surface have been established or depend on solubility behaviour or desorption conditions. These uncertainties relating to the conditions of impregnation can lead to anomalies. For example a local excess of released BMP may activate bone resorption instead of bone formation[43]. In contrast, adsorption leads to stable association, and the molecules can only be displaced by mineral ions and/or soluble proteins with a stronger affinity for apatite surfaces, but in a predicable manner. In general, spontaneous release is very low and the active molecules show sustained release related to the degradation of the biomaterial. It is probable, although this has not been systematically investigated, that the release rate is related to the nature of the apatite samples and their maturation stage. Another interesting property of adsorbed growth factors is their availability for cells, as demonstrated in the case of VEGF. In vivo studies have confirmed the efficiency and reproducibility of adsorbed BMP-2 in osteoinduction.

Cell adhesion on an apatite surface appears to be generally poorer than on specially treated surfaces like those of cell culture dishes. Therefore, several attempts have been made to favour cell attachment by using cell-binding proteins or specific protein sequences, such as RGD (arginine, glycine, aspartic acid), that are involved in cell attachment. The results generally show an improvement in cell adhesion and differentiation. It has been suggested that the osteoconductive properties of apatite could be due to the ability of its surface to bind such proteins.

CONCLUSION

This non-exhaustive review of some adsorption studies, essentially carried out by our research groups in recent years, demonstrates the progressive shift of our conception of the adsorption process from an electrostatic model based on previous considerations, to a more focused ion exchange model. The possibilities of interactions between a molecule, especially a macromolecule, and a mineral apatite surface are of course more numerous than the simple ion exchange reactions proposed here, but our experience suggests that this is a fundamental basic reaction determining the strong adsorptive properties of many molecules and bio-macromolecules of interest and explaining some specific behaviours like the effect of phosphate and calcium ions on the adsorption equilibrium and the apparent irreversibility following dilution. The observation of such ion exchange interactions needs a precise evaluation of the amount adsorbed and of the mineral ion content of the solution, which is rarely carried out in most adsorption studies. Among the competitive reactions that have to be taken into account we have identified the apatite dissolution equilibrium and the surface equilibration reactions of the mineral in solution, which essentially involves the surface hydrolysis of phosphate species associated with the release of calcium ions. These different equilibria can lead to different but totally equivalent descriptions of the adsorption phenomena.

REFERENCES

[1]T. Matsumoto, M. Okazaki, M. Inoue, S. Yamaguchi, T. Kusunose, T. Toyonaga, Y. Hamada, J. Takahashi, Hydroxyapatite particles as a controlled release carrier of protein, *Biomaterials*, **25(17)**, 3807-3812 (2004).
[2]C.J. Wilson, R.E. Cleqq, D.I. Leavesley, M.J. Pearcy, Mediation of biomaterial–cell interactions by adsorbed proteins: a review, *Tissue Eng.*, **11**, 1-18 (2005).
[3]J.E. Davies, In vitro modeling of the bone/implant interface, *Anat. Rec.*, **245(2)**, 426-445 (1996).
[4]A.L. Boskey, W. Ullrich, L. Spevak, H. Gilder, Persistence of complexed acidic phospholipids in rapidly mineralizing tissues is due to affinity for mineral and resistance to hydrolytic attack: in vitro data, *Calcif. Tissue Int.*, **58(1)**, 45-51 (1996).
[5]P.V. Hauschka, F.H. Wians, Osteocalcin–hydroxyapatite interactions in the extracellular organic matrix of bone, *Anat. Rec.* **224(2)**, 180-188 (1989).
[6]R.W. Romberg, P.G. Werness, B.L. Riggs, K.G. Mann, Inhibition of apatite crystal growth by bone-specific and other calcium-binding proteins, *Biochemistry,* **25**, 1176-1180 (1986).
[7]H.A. Goldberg, K.J. Warner, G.K. Hunter, Binding of bone sialoprotein osteopontin and synthetic polypeptides to hydroxyapatite, *Connect. Tissue Res.* **42**, 25-37 (2001).
[8]C. Robinson, S.J. Brookes, J. Kirkham, W.A. Bonass, R.C. Shore, Crystal growth in dental enamel: the role of amelogenins and albumin, *Adv. Dent. Res.,* **10(2)**, 173-180 (1996).
[9]A. Jung, S. Bisaz, H. Fleish, The binding of pyrophosphate and two diphosphonates by

hydroxyapatite crystals, *Calcif. Tissue Res.*, **11**, 269-280 (1973).

[10]C. Combes, C. Rey, Adsorption of proteins and calcium phosphate materials bioactivity, *Biomaterial*, **23**, 2817-2823 (2002).

[11]L. Addadi, J. Moradian-Oldak, H. Füredi-Milhofer, S. Weiner, A. Veis, Stereochemical aspects of crystal regulation in calcium phosphate-associated mineralized tissues. In: *Chemistry and biology of mineralized tissues*, Ed. H. Slavkin, Elsevier, 153-162 (1992).

[12]V. Hlady, J. Buijs, Protein adsorption on solid surfaces, *Current Opinion in Biotechnology*, **7**, 72-77 (1996).

[13]A.J. Garcia, B.G. Keselowsky, Biomimetic surfaces for control of cell adhesion to facilitate bone formation, *Crit. Rev. Eukaryot Gene Expr.*,**12(2)**, 151-162 (2002).

[14]G.H. Nancollas, R. Tang, R.J. Phips, Z. Henneman, S. Gulde, W. Wu, A. Mangood, R.G.G. Russel, F.H. Ebetino, Novel insights into actions of bisphosphonate on bone: differences in interactions with hydroxyapatite, *Bone*, **38**, 617-627 (2006).

[15]S. Hjerten, O. Levin, A. Tiselius, Protein chromatography on calcium phosphate columns, *Arch. Biochem. Biophys.*, **65**, 132-155 (1956).

[16]M. Rouahi, E. Champion, O. Gallet, A. Jada, K. Anselme, Physico-chemical characteristics and protein adsorption potential of hydroxyapatite particles: influence on in vitro biocompatibility of ceramics after sintering. *Colloids and surfaces B: Biointerfaces*, **47**, 10-19 (2006).

[17]G. Yin, Z. Liu, J. Zhan, F. Ding, N. Yuan, Impact of the surface charge property on protein adsorption on hydroxyapatite, *Chem. Eng. J.* **87**, 181-186 (2002).

[18]C. Rey, C. Combes, C. Drouet, A. Lebugle, H. Sfihi, A. Barroug, Nanocrystalline apatites in biological systems, characterization, structure and properties, *Mat. Wiss. u. Werstofftech.* **38**, 996-1002 (2007).

[19]A. Barroug, L.T. Kuhn, L.C. Gerstenfeld, M.J. Glimcher, Hydroxyapatite crystals as a local delivery system for cisplatin: adsorption and release of cisplatin in vitro, *J. Orthopaedic Research*, **22(4)**, 703-708 (2004).

[20]S. Ouizat, A. Barroug, A. Legrouri, C. Rey, Adsorption of bovine serum albumin on poorly crystalline apatite: influence of maturation, *Materials Research Bulletin* **34(14/15)**, 2279-2289 (1999).

[21]L. Benaziz, A. Barroug, A. Legrouri, C. Rey, A. Lebugle, Adsorption of o-phospho-L-serine and L-serine onto poorly crystalline apatite. *J. Colloid Interf. Sci.*, **238**, 48-53 (2001).

[22]A. Barroug, A. Legrouri, C. Rey, Exchange reactions at calcium phosphate surface and applications to biomaterials, *Key Engin. Mater.*, **361-363**, 79-82 (2008).

[23]Y. Wu, J.L. Ackerman, E.S. Strawich, C. Rey, H.M. Kim, M.J. Glimcher, Phosphate ions in bone: identification of a calcium phosphate complex by ^{31}P solid-state NMR spectroscopy at early stages of mineralization, *Calcif. Tissue Int.*, **72**, 610-626 (2003).

[24]M. Subirade, A. Lebugle, Study of the mineral–organic interface between a calcium phosphate and a phosphorylated monomer, *Ann. Chim. Fr.*, **16**, 183-193 (1993).

[25]S. Josse, C. Faucheux, A. Soueidan, G. Grimandi, D. Massiot, B. Alonso, P. Janvier, S. Laïb, P. Pilet, O. Gauthier, G. Daculsi, J.J. Guicheux, B. Bujoli, J.M. Bouler, Novel biomaterials for bisphosphonate delivery, *Biomaterials*, **26(14)**, 2073-2080 (2005).

[26]K. Ohta, H. Monma, S. Takahashi, Adsorption characteristics of proteins on calcium phosphates using liquid chromatography, *J. Biomed. Mater. Res.*, **55(3)**, 409-414 (2001).

[27]D.T.H. Wassel, R.C. Hall, G. Embery, Adsorption of bovine serum albumin onto hydroxyapatite, *Biomaterials*, **16**, 697-702 (1995).

[28]A.P. Serro, M. Bastos, J. Costa Pessoa, B. Saramago, Bovine serum albumin conformational changes upon adsorption on titania and on hydroxyapatite and their relation with biomineralization, *J. Biomed. Mater. Res.*, **70A**, 420-427 (2004).

[29]S. Cazalbou, C. Combes, D. Eichert, C. Rey, M.J. Glimcher, Poorly crystalline apatites: evolution and maturation in vitro and in vivo, *J. Bone Miner. Metab.*, **22(4)**, 310-317 (2004).

[30]T. Boix, J. Gomez-Morales, J. Torrent-Burguès, A. Monfort, P. Puigdomenech, R. Rodriguez-Clemente, Adsorption of recombinant human bone morphogenetic protein rhBMP-2m onto hydroxyapatite, *J. Inorg. Biochem.*, **99(5)**, 1043-1050 (2005).

[31]H. Autefage, F. Briand-Mesange, S. Cazalbou, C. Drouet, D. Fourmy, S. Gonçalves, J-P. Salles, C. Combes, P. Swider, C. Rey, Adsorption and release of BMP-2 on nanocrystalline apatite-coated and uncoated hydroxyapatite/b-tricalcium phosphate porous ceramics, *J. Biomed. Mater. Res.* (accepted).

[32]V. Midy, C. Rey, E. Brès, M. Dard, Basic fibroblast growth factor adsorption and release properties of calcium phosphate, *J. Biomed. Mater. Res.*, **41(3)**, 405-411 (1998).

[33]V. Midy, E. Hollande, C. Rey, M. Dard, J. Plouet, Adsorption of vascular endothelial growth factor to two different apatitic materials and its release, *J. Mater. Sci. Mater Med.*, **12(4)**, 293-298 (2001).

[34]C. Rill, Z.I. Kolar, G. Kickelbick, H.Th. Wolterbeek, J.A. Peters, Kinetic and thermodynamics of adsorption on hydroxyapatite of the [160]Tb terbium complexes of the bone-targeting ligand DOTP and BPPED. *Langmuir*, **25**, 2294-2301 (2009).

[35]A. Barroug, J. Fastrez, J. Lemaitre, P. Rouxhet, Adsorption of succinylated lysozyme on hydroxyapatite, *J. Coll. Interface Sci.*, **189**, 37-42 (1997).

[36]A. Barroug, E. Lernoux, J. Lemaitre, P.G. Rouxhet, Adsorption of catalase on hydroxyapatite, *J. Coll. Interface Sci.*, **208**, 147-152 (1998).

[37]E.I. Pearce, Ion displacement following the adsorption of anionic macromolecules on hydroxyapatite, *Calcif. Tissue Int.* **10**, 123-131 (1981).

[38]D.N. Misra, Interaction of some alkali metal citrates with hydroxyapatite. Ion-exchange adsorption and role of charge balance, *Colloids and Surfaces*, **141**, 173-179 (1998).

[39]A. Chhettry, Z. Wang, J. Hsu, J.L. Fox, A.A. Baig, A.M. Barry, H. Zhuang, M. Otsuka, W.I. Huguchi, Metastable equilibrium solubility distribution of carbonated apatite as a function of solution composition, *J. Coll. Interface Sci.*, **218**, 57-67 (1999).

[40]W. Wu, G.H. Nancollas, Determination of interfacial tension from crystallization and dissolution data: a comparison with other methods, *Adv. Coll. Interface Sci.*, **79**, 229-279 (1999).

[41]A. Boskey, Matrix proteins and mineralization: an overview, *Connect. Tissue Res.*, **35**, 357-363 (1996).

[42]M.J. Glimcher, The possible role of collagen fibrils and collagen phosphoprotein complexes in the calcification of bone in vitro and in vivo, *Biomaterials*, **11**, 7-10 (1990).

[43]J.M. Toth, S.D. Boden, J.K. Burkus, J.M. Badura, S.M. Peckham, W.F. McKay, Short-term osteoclastic activity induced by locally high concentrations of recombinant human bone morphogenetic protein-2 in a cancellous bone environment. *Spine*, **34**, 539-550 (2009).

BIOACTIVE CALCIUM PHOSPHATES AND NANOCOMPOSITE SCAFFOLDS FOR BONE TISSUE ENGINEERING

Min Wang*
Department of Mechanical Engineering, The University of Hong Kong
Pokfulam Road, Hong Kong
Email: memwang@hku.hk

ABSTRACT

For their use in human body tissue repair, synthetic bioactive calcium phosphates (Ca-P, including bioactive apatites), which are mainly developed for hard tissue replacement, have been under intensive as well as extensive investigations worldwide over the past three decades. Tissue engineering, as a promising means for human tissue repair, emerged two decades ago. Tissue engineering requires the use of (different) cells and the provision initially of an appropriate microenvironment for the cells to attach and proliferate, leading to the formation of new tissue. In scaffold-based bone tissue engineering, biodegradable, three-dimensional porous scaffolds are a key requirement. Synthetic hydroxyapatite or bioactive Ca-P nanoparticles can be incorporated in biodegradable polymers to form scaffolds, imparting scaffolds with osteoconductivity and hence enhancing bone formation. Processes have been developed for synthesizing Ca-P nanoparticles and various methods have been employed to fabricate nanocomposite scaffolds for bone tissue engineering. The influencing factors were also studied.

INTRODUCTION

In the development of biomaterials for various purposes, the research on bioceramics such as carbon, alumina and zirconia ceramics, which are "bioinert", has a relatively long history. (In this paper, "bioceramics" is used broadly to also include bioactive glasses and glass-ceramics.) For their use in human body tissue repair, synthetic bioactive apatites and calcium phosphates (Ca-P) have been under intensive as well as extensive investigations worldwide for over three decades now. The extensive literature on the R & D of bioactive apatites and calcium phosphates has, to a large extent, systematically documented the research efforts and achievements by various research groups around the world [1-7]. This group of bioactive bioceramics has been traditionally viewed as materials suitable for hard tissue replacement as in most cases they are perceived as non-biodegradable. It may be right to (first of all, believe, and then) state that the branch of bioactive apatites and calcium phosphates of the bioceramics family has now matured and that the continued research in this area is and will be mainly applications-oriented.

Tissue engineering, as a new, promising means for human body tissue repair, emerged just over two decades ago [8, 9] and has been attracting great attention in science, engineering, medicine, and the society [10, 11]. The rapid development in this area is evidenced by the large and also increasing number of publications in recent years dealing with various aspects of engineering body tissues as well as frequent international conferences on tissue engineering. Governments of industrialized countries as well as some developing countries have been providing large fundings for research in this area and the recent lift of ban on the US federal funding for stem cell research is obviously a boost to the R & D of tissue engineering. The emphasis in the tissue engineering route for human tissue repair is (natural) tissue regeneration, whose process requires the use of (different) cells and the provision initially of an appropriate microenvironment for the cells to attach and proliferate, leading to the eventual formation of the new tissue. In scaffold-based bone tissue engineering strategies, three-dimensional (3D) porous scaffolds are a key requirement because they can mimic the structure and function of extracellular matrix (ECM) of bone and support expanded cells to proliferate and secrete their own ECMs, thus replacing the biodegradable scaffolds with new bone.

In developing new biomaterials for various applications, the composite approach has been adopted by many researchers around the world. Increasingly, composite scaffolds for tissue engineering have been investigated and used. In world's premier biomaterials conferences, special symposia have been organized for biomedical composites [12] and composite tissue engineering scaffolds [13]. The osteoconductivity of synthetic apatites and calcium phosphates can be advantageously utilized through composite scaffolds for bone tissue engineering [14]. Therefore, various groups around the world have been conducting research in bioactive bioceramic-polymer composite scaffolds. This paper reviews some of our efforts in the development of composite scaffolds.

BIOACTIVE BIOCERAMICS AND CALCIUM PHOSPHATES

Bioactive bioceramics
With the exception of carbon, which has been used mainly in the form of pyrolytic carbon for mechanical heart valves due to its excellent hemocompatibility and other properties, ceramic biomaterials had not been seriously considered as viable biomaterials until the 1970s. Over the past 40 years, bioceramics, which include ceramics, glasses and glass-ceramics that are used for human hard tissue repair, have been widely investigated [15-17]. Bioceramics such as bioinert alumina and toughened zirconia ceramics are now used in implants for hip joints [18]. For nearly three decades since the early 1980s, extensive research has been conducted on bioactive bioceramics such as hydroxyapatite (HA, $Ca_{10}(PO_4)_6(OH)_2$), Bioglass® and A-W glass-ceramic because these materials can form chemical bonding with bone after implantation [1, 2, 4]. (In this paper, "bioactive" is used interchangeably with "osteoconductive".) However, most bioactive bioceramics including HA are weak ceramics and hence cannot be used on their own for load-bearing applications in human bodies.

Even though bulk bioactive glasses or apatites lack the mechanical strength that is required for load-bearing implants, their bioactivity can be and indeed have been utilized in two different ways:
(1)   forming a bioactive coating on metal implants;
(2)   being incorporated in polymers, metals or ceramics to form bioactive composites.
Bioactive single-component or composite coatings can be formed on metal implants [19]. A variety of coating techniques and different bioactive bioceramics can be used to form the bioactive coatings [20-27]. Even with several shortcomings, plasma spraying has been used as a standard industrial process to fabricate "hydroxyapatite"-coated hip implants.

Various non-porous bioactive composites have been produced and investigated for possible hard tissue replacement and in the case of biodegradable polymer matrices, tissue regeneration [14]. The bioactive bioceramics incorporated in these composites can be in different forms: particles, whiskers and fibers (short or long). Particulate hydroxyapatite [28], tricalcium phosphate [29], Bioglass® [30] and A-W glass-ceramic [31] were incorporated into "biostable" or biodegradable polymers to form various composites. These polymer-based bioactive composites exhibited various degrees of osteoconductivity and HAPEX^{TM}, the first bioactive ceramic-polymer composite that is designed to mimic the structure and match properties of bone, has been used for bone-substituting implants [32,33].

Bioactive apatites and calcium phosphates

Calcium phosphate-based bioceramics have been in use in dentistry and medicine for more than 20 years now. The interest in the prominent member of the calcium phosphate family, hydroxyapatite (HA), arises from its similarity to bone apatite, the major component of the inorganic phase of bone. The pioneering work by Jarcho, de Groot and Aoki on synthetic HA in the 1970s opened up the research and led to the rapid development of a major category of current biomaterials, bioactive bioceramics [4]. Other bioactive apatites, such as fluorapatite and chlorapatite, were synthesized and studied. Apart from HA, other bioactive calcium phosphates, such as calcium-deficient HA (CDHA) and tricalcium phosphate (TCP, $Ca_3(PO_4)_2$), were made and investigated for various purposes. Many review articles and book chapters give excellent accounts of the work on these materials [1-7,34]. The series of annual (becoming biennial from 2009) *Bioceramics* symposium under the auspices of the International Society for Ceramics in Medicine provides information on the latest research of these materials. However, as has been pointed out earlier, due to their brittleness and fragility, bioactive calcium phosphates on their own cannot be used for load-bearing implants.

Prior to the worldwide surge in nanomaterial and nanotechnology research, HA powders of the micrometer size had mainly been produced and used for various purposes: non-porous (non-load bearing) implants, stock materials for bioactive coatings, bioactive phase in non-porous biomedical composites, etc. There are several routes for the production of micro-size HA powders: precipitation method, hydrolysis method, solid-state reaction, hydrothermal reaction, etc. The wet-synthesis (precipitation method) has been the most commonly used method for HA mass production. One of the wet-synthesis processes used the precipitation reaction between calcium hydroxide and orthophosphoric acid to produce a stoichoimetric HA slurry [35]. The slurry was then spray-dried to form HA powders. The HA synthesized were spherical in shape and had the average diameter of ~17μm. They could be used for plasma spraying to form bioactive coatings on metal implants [20] and for making bioceramic-polymer composites for hard tissue replacement [36].

In recent years, nanometer-size particles of various calcium phosphates have been made using different methods. A nanoemulsion technique was developed to synthesize nanoparticles of carbonated HA [37] and Ca-P [38]. The shape and stoichoimetry of nanoparticles could by controlled by the reaction condition. For producing Ca-P nanoparticles, calcium nitrate tetrahydrate reacted with diammonium phosphate in a certain pH range. The precipitates were then freeze-dried. The Ca-P synthesized was spherical in shape and were 10-30nm in diameter. Both nano-size HA and Ca-P particles have been used for bone tissue engineering scaffolds [39-43].

NANOCOMPOSITE SCAFFOLDS FOR BONE TISSUE ENGINEERING

Biodegradable polymers for tissue engineering applications

Various materials have been investigated or used for making porous scaffolds for tissue engineering applications [44]. However, the majority of tissue engineering scaffolds that have gone into investigations of *in vivo* tissue synthesis so far are made dominantly by a few biodegradable polymers. Traditionally, polyesters such as poly(L-lactic acid) (PLLA) and poly(lactic acid-co-glycolic acid) (PLGA) are materials of choice for tissue engineering. Other synthetic biodegradable polymers such as poly(ε-caprolactone) (PCL) and poly(propylene fumarate) (PPF) have also been investigated for making tissue engineering scaffolds. Natural, biodegradable polymers such polyhydroxybutyrate (PHB, which is made by many microorganisms) and its copolymer poly(hydroxybutyrate-co-hydroxyvalerate) (PHBV), chitosan and collagen are also used in tissue engineering, having advantages over synthetic polymers because their degradation products are part of the body or body fluids. But natural polymers have their own problems for tissue engineering applications. The degradation rate of these polymers (natural or synthetic), which is affected by various factors but predominantly by the average molecular weight of the polymer, controls the degradation rate of scaffold. Recognizing the importance of controlling the degradation rate and other properties of scaffolds, polymer blends have also been of interest for constructing tissue engineering scaffolds [42]. But nearly all biocompatible and biodegradable polymers are considered to be non-osteoconductive.

It is important to bear in mind that due to the requirements of high porosity and interconnectivity of pores and also due to the material nature of polymers, polymer and polymer-based composite scaffolds are weak structures. Their strength decreases exponentially with an increasing degree of porosity [41]. The scaffolds on their own are not destined for load-bearing applications.

The composite approach in developing bone tissue engineering scaffolds

For bone tissue engineering, osteoconductivity of the scaffold is one of the major considerations. Bioactive bioceramic-polymer composite scaffolds, which take the advantage of osteoconductivity of the bioceramics, are promising structures for bone cell attachment and proliferation and tissue formation and hence are actively investigated by many research groups. It is also highly desirable that, if a composite scaffold is used for bone tissue engineering, the composite scaffold should be totally biodegradable.

Biodegradable polymer-based scaffolds containing bioactive apatites or calcium phosphates can be produced. In these scaffolds, the bioceramics serve two purposes: (i) making the scaffolds osteoconductive, and (ii) reinforcing the scaffolds. Generally, there are two approaches for making suitable bioceramic-polymer composite scaffolds:
(1)   incorporating bioceramic particles in the scaffold through a variety of techniques; and
(2)   coating a polymer scaffold with a thin layer of apatite through a biomimetic process.
These two approaches have respective advantages and disadvantages and the latter can be used to make the readily fabricated non-bioactive polymer scaffolds osteoconductive. Using either approach, there are a number of factors, such as matrix polymer, bioceramic phase, interactions between the polymer matrix and the bioceramic, etc., that must be considered in order for the tissue engineering strategy to be successful.

Composite scaffolds containing bioactive bioceramic particles

Incorporating bioceramic particles (microparticles previously, and nanoparticles in recent years) into biodegradable polymers to form bioactive scaffolds can be successful. But there is an upper limit for the amount of particles that can be incorporated. A variety of techniques including porogen leaching [45], emulsion freezing / freeze-drying [41, 42, 46, 47], electrospinning [39, 48, 49], and selective laser sintering (SLS) [40, 43, 50-54] have been /are investigated actively for scaffold fabrication using bioactive HA or Ca-P particles for the composite scaffolds. With optimized fabrication parameters, bioceramic nanoparticles were evenly distributed in composite scaffolds. The composite scaffolds exhibited higher compressive strength and modulus than scaffolds made of pure polymer or polymer blends. *In vitro* biodegradation tests indicated that these scaffolds were biodegradable and *in vitro* experiments using osteoblastic cell culture showed that these scaffolds were osteoconductive.

Scaffolds containing therapeutic agents, viz., drugs and growth factors such as bone morphogenetic protein (BMP), can enhance tissue regeneration. It has been shown that biomolecules could be incorporated in composite scaffolds made via SLS [43]. The model protein bovine serum albumin (BSA) incorporated in Ca-P/PHBV scaffolds could be released gradually. In comparison with pure PHBV scaffolds, the incorporation of Ca-P nanoparticles in composite scaffolds increased the BSA encapsulation efficiency and release rate. A mathematical model could be used to simulate the BSA release profiles from Ca-P/PHBV scaffolds.

Osteoconductive composite scaffolds could also be made using micro-size Bioglass® particles [55] or other particulate bioactive bioceramics.

Composite scaffolds possessing a bioactive bioceramic coating

Coating the pore surface of a polymer scaffold with a thin layer of apatite or apatite-containing composite layer is another effective approach in making osteoconductive scaffolds for bone tissue engineering. The coatings could be formed through biomimetic processes [56-62]. In the classic biomimetic process to form an apatite layer on metal or ceramic surfaces, a normal-strength simulated body fluid (SBF) is commonly used [63] and it usually takes 1-4 weeks to form the apatite layer. However, as polymeric materials such as PLLA are easily hydrolyzed in water, a much shorter coating time must be used for biomimetic deposition on polymer scaffolds. Therefore, accelerated biomimetic deposition processes which employ higher-strength SBFs (2SBF, 5SBF, etc.) can be used. With accelerated biomimetic deposition, an apatite layer could form on the surfaces of poly(glycolic acid) (PGA) fiber meshes and PLLA scaffolds within 24 hours [56-58] There were concerns that in the static biomimetic deposition processes, less apatite was formed in the interior of a scaffold than the peripheral region because of easier ions exchange between the peripheral region and the surrounding coating solution. The formation of a uniform spatial distribution of apatite coating in the scaffold (interior and surface areas) could be achieved by using dynamic biomimetic deposition [61]. With dynamic biomimetic deposition, a circulation was established, in which the coating solution flowed directly through the polymer scaffolds. Therefore, high transport of ions was achieved between the internal regions and flowing coating solution, and consequently more apatite particles were formed on the internal pore walls in the scaffold. It was also possible to fabricate apatite/collagen composite coatings on pore surfaces of PLLA scaffolds [56, 59]. The structure of apatite/collagen composite coatings mimicked that of natural bone. *In vitro* cell culture experiments revealed that PLLA scaffolds coated with a layer of either apatite or apatite/collagen composite promoted osteoblastic cell adhesion and proliferation [60, 62]. PLLA scaffolds with an apatite/collagen composite coating showed better improvement in osteoconductivity than PLLA scaffolds with an apatite coating.

Osteoconductive composite scaffolds could also be made by coating pure polymer scaffolds with Bioglass® particles [64].

CONCLUDING REMARKS

Bioactive calcium phosphates are used in non-porous polymer-based composites for hard tissue replacement/regeneration and in porous scaffolds for bone tissue engineering applications. HA and Ca-P nanoparticles can be synthesized through nanoemulsion processes. Tissue engineering scaffolds containing these nanoparticles can be produced using a variety of techniques. The bioactive bioceramics in composite scaffolds impart osteoconductivity on the scaffolds, making them suitable for bone tissue engineering. *In vitro* cell culture studies have shown enhanced osteoblastic cell adhesion and proliferation. Bioactive calcium phosphates play important roles in bone tissue repair.

ACKNOWLEDGEMENTS

I thank the students and research staff in Nanyang Technological University (NTU), Singapore, The Hong Kong Polytechnic University (PolyU) and The University of Hong Kong (HKU), Hong Kong, for working with me on our research reviewed in this paper. I also thank my collaborators in these and other universities. Assistance provided by technical staff in the universities is appreciated. I acknowledge the support provided by funding agencies (EPSRC in the UK, MOE in Singapore, and RGC in Hong Kong) and universities (University of London, NTU, PolyU and HKU).

REFERENCES

[1]K. de Groot, *Bioceramics of Calcium Phosphate*, CRC Press, Boca Raton, (1983).
[2]T. Yamamuro, L.L. Hench, J. Wilson, (Eds.), *Handbook of Bioactive Ceramics*, Vols.1 & 2, CRC Press, Boca Raton, (1990).
[3]R.Z. LeGeros, *Calcium Phosphates in Oral Biology and Medicine*, Karger, Basel, (1991).
[4] L.L. Hench, J. Wilson, (Eds.), *An Introduction to Bioceramics*, World Scientific, Singapore, (1993).
[5]J.C. Elliott, *Structure and Chemistry of the Apatites and other Calcium Orthophosphates*, Elsevier, Amsterdam, (1994).
[6]H. Aoki, *Medical Applications of Hydroxyapatite*, Isiyaku EuroAmerica, Tokyo, (1994).
[7]L.C. Chow, E.D. Eanes, (Eds.), *Octacalcium Phosphate*, Karger, Basel, (2001).
[8]R. Skalak, C.F. Fox, (Eds.), *Tissue Engineering*, Alan R.Liss Inc., New York, (1988).
[9]R. Langer, J.R. Vacanti, Tissue Engineering, *Science*, **260**, 920-926, (1993).
[10]L.V. McIntire, (Ed.), *WTEC Panel on Tissue Engineering Research: Final Report*, Academic Press, San Diego, (2003).
[11]M.J. Lysaght, A.L. Hazlehurst, Tissue Engineering: The End of the Beginning, *Tissue Engineering*, **10**, 309-320, (2004).
[12]The 7th World Biomaterials Congress, Symposium on Developing New Biomaterials: the Composite Approach, Symposium Organized by Min Wang and Akiyoshi Osaka, Sydney, May 2004.
[13]The 8th World Biomaterials Congress, Symposium on Composite Scaffolds for Tissue Engineering, Symposium Organized by Min Wang and Jan Stegemann, Amsterdam, The Netherlands, May 2008.
[14] M. Wang, Developing Bioactive Composite Materials for Tissue Replacement, *Biomaterials*, **24**, 2133-2151, (2003).
[15]G.D. Winter, D.F. Gibbons, H. Plenk Jr., (Eds.), *Biomaterials 1980*, John Wiley, Chichester, (1982).
[16]H. Oonishi, H. Aoki, K. Sawai, (Eds.), *Bioceramics*, Vol.1, Ishiyaku EuroAmerica, Tokyo, (1989).
[17]M. Prado, C. Zavaglia, (Eds.), *Bioceramics*, Vol.21, Trans Tech Publications, Switzerland, (2009).
[18]M. Wang, "Bioceramics", in: *Recent Research Developments in Biomaterials*, Edited by Y. Ikada,

Research Signpost, Trivandrum, 33-76, (2002).

[19]M. Wang, "Composite Coatings for Implants and Tissue Engineering Scaffolds", in: *Biomedical Composites*, Edited by L. Ambrosio, Woodhead Publishing, Cambridge, UK, in press, (2009).

[20] M. Wang, X.Y. Yang, K.A. Khor, Y. Wang, Preparation and Characterisation of Bioactive Monolayer and Functionally Graded Coatings, *J of Materials Science: Materials in Medicine*, **10**, 269-273, (1999).

[21]C.X. Wang, Z.Q. Chen, M. Wang, Z.Y. Liu, P.L. Wang, Ion-beam-sputtering/mixing Deposition of Calcium Phosphate Coatings: I. Effects of Ion Mixing Beams, *Journal of Biomedical Materials Research*, **55**, 587-595, (2001).

[22]R.R. Kumar, M. Wang, Modulus and Hardness Evaluations of Sintered Bioceramic Powders and Functionally Graded Bioactive Composites by Nano-indentation Technique, *Materials Science and Engineering A: Structural materials: properties, microstructure and processing*, **338**, 230-236, (2002).

[23]C.X. Wang, M. Wang, X. Zhou, Nucleation and Growth of Apatite on Chemically Treated Titanium Alloy: An Electrochemical Impedance Spectroscopy Study, *Biomaterials*, **24**, 3069-3077, (2003).

[24]J.M. Wu, M. Wang, Y.W. Li, F.D. Zhao, X.J. Ding, A. Osaka, Crystallization of Amorphous Titania Gel by Hot Water Aging and Induction of *In Vitro* Apatite Formation by Crystallized Titania, *Surface & Coatings Technology*, **201**, 755-761, (2006).

[25]T. Sun T, M. Wang, Low-temperature Biomimetic Formation of Apatite/TiO_2 Composite Coatings on Ti and NiTi Shape Memory Alloy and their Characterization, *Applied Surface Science*, **255**, 396-400, (2008).

[26]T. Sun, J.H.W. Wong, M. Wang, Fabrication of an Apatite/Collagen Composite Coating on the NiTi Shape Memory Alloy through Electrochemical Deposition and Coating Characterisation, *Materials Science Forum*, **618-619**, 319-323, (2009).

[27]T. Sun, L.P. Wang, M. Wang, Fabrication and Characterization of Ti-O/Ti Composite Coatings on NiTi Shape Memory Alloy for Medical Applications, *Proceedings of WACBE World Congress on Bioengineering 2009*, Hong Kong, 2009.

[28]M. Wang, D. Porter, W. Bonfield, Processing, Characterisation, and Evaluation of Hydroxyapatite Reinforced Polyethylene Composites, *British Ceramic Transactions*, **93**, 91-95, (1994).

[29]L.J. Chen, M. Wang, Production and Evaluation of Biodegradable Composites Based on PHB-PHV Copolymer, *Biomaterials*, **23**, 2631-2639, (2002).

[30]M. Wang, L.L. Hench, W. Bonfield, Bioglass®/High Density Polyethylene Composites for Soft Tissue Applications: Preparation and Evaluation, *Journal of Biomedical Materials Research*, **42**, 577-586, (1998).

[31]M. Wang, T. Kokubo, W. Bonfield, A-W Glass-Ceramic Reinforced Polyethylene Composite for Medical Applications, *Bioceramics*, Vol.9, 387-390, (1996).

[32]R.N. Downs, S. Vardy, K.E. Tanner, W. Bonfield, Hydroxyapatite-Polyethylene Composite in Orbital Surgery, *Bioceramics*, Vol.4, 239-246, (1991).

[33]R.E. Swain, M. Wang, B. Beale, W. Bonfield, HAPEX™ for Otologic Applications, *Biomedical Engineering: Applications, Basis & Communications*, **11**, 315-320, (1999).

[34]L.L. Hench, Bioceramics: From Concept to Clinic, *Journal of the American Ceramic Society*, **74**, 1487-1510, (1991).

[35]H.Y. Yang, M. Wang, Effects of Reaction Parameters on the Thermostability of Spray-Dried Hydroxyapatite Powders, *Proceedings of the 8th International Conference on Processing and Fabrication of Advanced Materials*, Singapore, 1999, 307-316.

[36]M. Wang, C.Y. Yue, B. Chua, Production and Evaluation of Hydroxyapatite Reinforced Polysulfone for Tissue Replacement, *Journal of Materials Science: Materials in Medicine*, **12**, 821-826, (2001).

[37]W.Y. Zhou, M. Wang, W.L. Cheung, B.C. Gao, D.M. Jia, Synthesis of Carbonated Hydroxyapatite Nanospheres through Nanoemulsion, *Journal of Materials Science: Materials in Medicine*, **19**, 103-110, (2008).

[38]B. Duan, M. Wang, W.Y. Zhou, W.L. Cheung, Synthesis of Ca-P Nanoparticles and Fabrication of Ca-P/PHBV Microspheres for Bone Tissue Engineering Applications, *Applied Surface Science*, **255**, 529-533, (2008).

[39]H.W. Tong, M. Wang, Electrospinning of Aligned Biodegradable Polymer Fibers and Composite Fibers for Tissue Engineering Applications, *Journal of Nanoscience and Nanotechnology*, **7**, 3834-3840, (2007).

[40]W.Y. Zhou, S.H. Lee, M. Wang, W.L. Cheung, W.Y. Ip, Selective Laser Sintering of Porous Tissue Engineering Scaffolds from Poly(L-Lactide)/Carbonated Hydroxyapatite Nanocomposite Microspheres, *Journal of Materials Science: Materials in Medicine*, **19**, 2535-2540, (2008).

[41]N. Sultana, M. Wang, Fabrication of HA/PHBV Composite Scaffolds through the Emulsion Freezing / Freeze-drying Process and Characterisation of the Scaffolds, *Journal of Materials Science: Materials in Medicine*, **19**, 2555-2561, (2008).

[42]N. Sultana, M. Wang, PHBV/PLLA-based Composite Scaffolds Containing Nano-sized Calcium Phosphate Particles for Bone Tissue Engineering, *J of Experimental Nanoscience*, **3**, 121-132, (2008).

[43]B. Duan, M. Wang, Incorporation and Release of Biomolecules from Ca-P/PHBV Nanocomposite Tissue Engineering Scaffolds, *Proceedings of the 17th International Conference on Processing and Fabrication of Advanced Materials*, Delhi, India, 2008, 881-891.

[44]R.P. Lanza, R. Langer, J. Vacanti, (Eds.), *Principles of Tissue Engineering*, 2nd Edn. Academic Press, San Diego, (2000).

[45]J. Weng, M. Wang, J. Chen, Plasma Sprayed Calcium Phosphate Particles with High Bioactivity and their Use in Bioactive Scaffolds, *Biomaterials*, **23**, 2623-2629, (2002).

[46]N. Sultana, M. Wang, PHB- and PHBV-based Osteoconductive Composite Scaffolds Produced through an Emulsion Freeze-drying Process, *Proceedings of the 10th Annual Conference on Ceramics, Cells and Tissues*, Faenza, Italy, 2006, 201-208.

[47]N. Sultana, M. Wang, Fabrication and Characterisation of Polymer and Composite Scaffolds Based on Polyhydroxybutyrate and Poly(hydroxybutyrate-co-hydroxyvalerate), *Key Engineering Materials*, **334-335**, 1229-1232, (2007).

[48]H.W. Tong, M. Wang, PHBV-based Composite Nanofibers for Bone Tissue Engineering, *Proceedings of the 10th Annual Conference on Ceramics, Cells and Tissues*, Faenza, Italy, 2006,192-200.

[49]H.W. Tong, M. Wang, Biodegradable Microfibers Containing Hydroxyapatite Nanospheres for Bone Tissue Engineering, *Transactions of the Society For Biomaterials 2007 Annual Meeting*, Chicago, IL, USA, 2007, p583.

[50]W.Y. Zhou, W.L. Cheung, M. Wang, Biodegradable Composite Scaffolds Produced by Selective Laser Sintering, *Transactions of the Society For Biomaterials 2007 Annual Meeting*, Chicago, IL, USA, 2007, p170.

[51]W.Y. Zhou, M. Wang, W.L. Cheung, C.K. Yuen, W.Y. Ip, Selective Laser Sintered Meso-porous PLLA and PLLA/CHAp Nanocomposite Scaffolds for Bone Tissue Engineering, *Proceedings of the 8th World Biomaterials Congress*, Amsterdam, The Netherlands, 2008, Paper #3172.

[52]B. Duan, M. Wang, W.Y. Zhou, W.L. Cheung, Totally Bioresorbable Tissue Engineering Scaffolds Based on Ca-P/PHBV Nanocomposite and Fabricated via Selective Laser Sintering: Part II. Scaffold fabrication and characterization, *Proceedings of the 8th World Biomaterials Congress*, Amsterdam, The Netherlands, 2008, Paper #1153.

[53]B. Duan, M. Wang, W.Y. Zhou, W.L. Cheung, Totally Bioresorbable Tissue Engineering Scaffolds Based on Ca-P/PHBV Nanocomposite and Fabricated via Selective Laser Sintering: Part III. Process optimization, *Proceedings of the 8th World Biomaterials Congress*, Amsterdam, The Netherlands, 2008, Paper #1309.

[54] B. Duan, M. Wang, W.Y. Zhou, W.L. Cheung, Design and Characteristics of Selective Laser Sintered Ca-P/PHBV Nanocomposite Scaffolds for Bone Tissue Engineering, *Proc. of the 4th International Symposium on Apatite and Correlative Biomaterials*, Manila, The Philippines, 2008, 75-79.

[55]A.R. Boccaccini, I. Notingher, V. Maquet, R. Jerome, Bioresorbable and Bioactive Composite Materials Based on Polylactide Foams Filled with and Coated by Bioglass® Particles for Tissue Engineering Applications, *Journal of Materials Science: Materials in Medicine*, 14, 443-450, (2003).

[56]Y. Chen, A.F.T. Mak, M. Wang, Formation of Apatite/Collagen Composite Coating on Poly(L-lactic acid) Scaffolds in an Accelerated Biomimetic Process, *Transactions of the Society For Biomaterials 30th Annual Meeting*, Memphis, TN, USA, 2005, p442.

[57]Y. Chen, A.F.T. Mak, M. Wang, Accelerated Formation of Bone-like Apatite on Biodegradable Polymer Substrates, *Key Engineering Materials*, **284-286**, 509-512, (2005).

[58]Y. Chen, A.F.T. Mak, J. Li, M. Wang, A.W.T. Shum, Formation of Apatite on Poly(α-hydroxy acid) in an Accelerated Biomimetic Process, *Journal of Biomedical Materials Research Part B: Applied Biomaterials*, **73B**, 68-76, (2005).

[59]Y. Chen, A.F.T. Mak, M. Wang, J. Li, Composite Coating of Bone-like Apatite Particles and Collagen Fibers on Poly L-lactic Acid Formed through an Accelerated Biomimetic Coprecipitation Process, *Journal of Biomedical Materials Research Part B: Applied Biomaterials*, **77B**, 315-322, (2006).

[60]Y. Chen, A.F.T. Mak, M. Wang, J. Li, M.S. Wong, PLLA Scaffolds with Biomimetic Apatite Coating and Biomimetic Apatite/collagen Composite Coating to Enhance Osteoblast-like Cells Attachment and Activity, *Surface & Coatings Technology*, **201**, 575-580, (2006).

[61]Y. Chen, A.F.T. Mak, M. Wang, Formation of Apatite within Biodegradable Scaffolds by an Accelerated Biomimetic Process in the Shaking Condition and Flow Condition, *Key Engineering Materials*, **334-335**, 1213-1216, (2007).

[62]Y. Chen, A.F.T. Mak, M. Wang, J. Li, M.S. Wong, *In vitro* Behaviour of Osteoblast-like Cells on PLLA Films with a Biomimetic Apatite or Apatite/collagen Composite Coating, *Journal of Materials Science: Materials in Medicine*, 19, 2261-2268, (2008).

[63]C. Ohtsuki, H. Kushitani, T. Kokubo, S. Kotani, T. Yamamuro, Apatite formation on the surface of Ceravital-type glass-ceramic in the body, *Journal of Biomedical Materials Research*, **25**, 1363–1370, (1991).

[64]A. Stamboulis, L.L. Hench, Bioresorbable Polymers: Their Potential as Scaffolds for Bioglass® Composites, *Key Engineering Materials*, **192-195**, 729-732, (2001).

PREPARATION OF HYDROXYAPATITE NANOPARTICLES APPLYING THE MICRO CHEMICAL PROCESS

Eiji Fujii[1], Koji Kawabata[1], Yoshiaki Nakazaki[2], Yuji Tanizawa[2], Akihiro Matsumoto[3], Yuki Shirosaki[3], Satoshi Hayakawa[3], Akiyoshi Osaka[3]

[1] Industrial Technology Center of Okayama Prefecture, 5301 Haga, Kita-ku, Okayama-shi, Japan 701-1296
[2] Nano Cube Japan Co. Ltd., 5303 Haga, Kita-ku, Okayama-shi, Japan 701-1296
[3] Okayama University, 3-1-1 Tsushima-naka, Kita-ku, Okayama-shi, Japan 700-8530

ABSTRACT

A micro-chemical process was applied to prepare nanometer-size hydroxyapatite (HA) particles with various shapes and sizes so as to provide high surface activities, where solutions of reagent-grade calcium nitrate tetrahydrate and diammonium hydrogen phosphate were introduced in a micro reactor. The particles were characterized by X-ray diffractometry, transmission electron microscopy, or inductively coupled plasma photometry. Particle size and shape were dependent on not only pH and concentration of the mixing solutions but also the flow rate. Dilute solutions containing 3 mM Ca(II) and 5 mM P(V) yielded spherical HA particles, whose average particle size increased from about 3 nm to 40 nm as the flow rate increased from 0.3 to 3 ml·min^{-1}, where pH and the temperature of the reaction system were respectively maintained at 10 and at 40°C. Concentrated solutions with 50 mM Ca(II) and 30 mM P(V) yielded rectangular plate-like HA particles, 300 nm long and 50 nm wide, when pH was kept at 8, while agglomerations of smaller HA particles of about 20-50 nm were obtained when pH was kept at 10.

INTRODUCTION

Hydroxyapatite (HA; $Ca_5(PO_4)_3OH$), a major component of bone mineral[1, 2], is commonly used clinically as a hard tissue-repairing material. Since the 1950s, much attention has been focused on its other applications: column packing material for chromatography to separate proteins or enzymes[3, 4, 5] and for removing pathogenic proteins from blood in blood purification therapy[6, 7, 8] as well as an adsorbent for albumin, viruses, and cells since Zittle et al. reported protein adsorption onto calcium phosphates in 1951[9]. HA accommodates most inorganic cations and anions by replacing the constituent calcium, phosphate, and hydroxyl ions (they include monovalent or divalent metallic cations, and multivalent oxoanions). Such ion substitution not only modifies lattice structures but also affects physical and chemical properties. For example, the introduction of Zn(II), Mg(II) and Si(IV) into apatite particles controlled their protein adsorption behavior[10, 11, 12].

Such adsorption-related properties as well as catalytic ability are associated with active surfaces. Thus, finely (nm-range) pulverized materials are favorable because the particles have a basically large specific surface area, and because they have large fractions of surface atoms that are chemically or physically active, and those fractions increase with a decrease in particle size.

185

Suppose a surface layer of 10 nm in depth is more chemically active than the core region for a particle of 100 nm in diameter. Then, the volume fraction is as large as 30 %. Therefore, nano-particles, or particles of nm size, are naturally superior in adsorbing ability. HA nano-particles are suitable for medical applications like drug-delivery systems (DDSs).

In this study, a novel micro-chemical process was applied to prepare nano-size HA particles with various shapes. Calcium and phosphate solutions are led to a tube-type micro reactor to be mixed together and precipitate HA particles—an inexpensive and easy procedure. The obtained HA samples were characterized by transmission electron microscopy (TEM), X-ray diffractometry (XRD) and inductively coupled plasma emission spectroscopy (ICP).

MATERIALS AND METHODS

Reagent-grade calcium nitrate tetrahydrate and diammonium hydrogen phosphate (Nacalai Tesque, Japan) solutions were dissolved into ultra-pure water to give two sets of solutions different in concentration for dilute and concentrated solution systems. For the former system, or the dilute system, we used a 5 mM calcium solution, and a 3 mM phosphate solution whose pH was adjusted to 10 by adding a 28 mass% NH_4OH aqueous solution. In contrast, for the concentrated solution system we used 50 mM calcium solution and 30 mM phosphate solutions whose pH was adjusted to 8, 9, and 10, as in the dilute solution system. These solutions, whose temperatures were kept at 40°C, were passed into a micro-reactor consisting of a 1mmϕ micro-tube, under a constant pressure of 0.2 MPa. Considering the pump capability, the flow rates employed for the dilute solution system were 0.3, 1.0, or 3.0 ml·min^{-1}, while a fixed rate of 3.0 ml·min^{-1} was used for the concentrated solution system. The flow rates were all in the region of turbulent flow and, hence, the calcium and phosphate solutions were subjected to vigorous mixing in the reactor. The out-coming suspension was collected and the HA particles were separated by filtration.

The obtained precipitates were dried at 105°C for 24 h in air and the Ca and P content of the samples were analyzed by inductively coupled plasma emission spectroscopy (ICP, iCAP6000, ThermoFisher). The crystalline phases were identified by an X-ray diffractometer (XRD, RINT2500, RIGAKU; CuKα, 40 kV-200 mA). The morphology of those sample particles was observed under a transmission electron microscopy (TEM, JEM-2100, JEOL).

RESULTS and DISCUSSION

(i) HA nano-particles from the dilute solution system

The as-dried samples derived from the dilute solution system, or from the 5 mM calcium nitrate tetrahydrate solution and the 3 mM diammonium hydrogen phosphate solution were coded as A, B, and C, depending on the flow rate at 0.3, 1.0 and 3.0 ml·min^{-1}. Their Ca/P ratios, derived from the ICP measurement, were 1.71, 1.68, and 1.69 for samples A, B, and C, respectively, and they were mostly stoichiometric. Figure 1 shows TEM micrographs of samples A, B and C. The average particle size of samples A and B were about 3 nm and 40 nm, respectively, and all of them were spherical in shape. In sample C, the component particles, 30 nm in size, were highly connected with each other and particle of sizes of more than 100 nm were formed. Thus, the particle size from the dilute solution system increased by increasing the flow rate. In sample A, homogeneity of particle size was excellent.

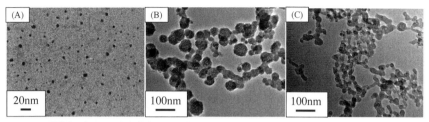

Figure 1. TEM micrographs of the samples, prepared from the dilute solution system (5 mM $Ca(NO_3)_2 \cdot 4H_2O$ and 3 mM $(NH_4)_2HPO_4$; pH 10; 40°C). Flow rate: (A) 0.3 ml·min^{-1}, (B) 1 ml·min^{-1}, (C) 3 ml·min^{-1}

Figure 2 shows the XRD patterns of samples B and C. Sample A, derived under a flow rate of 0.3 ml·min^{-1}, gave no diffractions, and the out-coming solution was transparent, though the TEM image in figure 1 (A) showed the presence of particles. All diffractions for samples B and C were assigned to those of hydroxyapatite (JCPDS 9-423)[13], and no secondary phases, such as calcite $(CaCO_3)$, octacalcium phosphate $(Ca_4H(PO_4)_3 \cdot 5H_2O)$, or brushite $(CaPO_3(OH) \cdot 2H_2O)$, were observed. Since samples B and C gave very similar XRD profiles (sharpness or height of each peak), the flow rate controlled the shape of the particles but not their crystallinity. Screwed particles were obtained for pH 8 and 9.

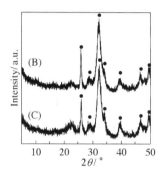

Figure 2. XRD patterns of the samples prepared from the dilute solution system, dried at 100°C for 24 h. Flow rate: (B) 1 ml·min^{-1}, (C) 3 m l·min^{-1}, •: Hydroxyapatite

(ii) HA nano-sheets from the concentrated solution system

Figure 3 shows TEM micrographs of the samples from the concentrated solution system, coded as D, E, and F according to the pH of the system, 8, 9 and 10, respectively. The morphology of sample F can be compared with that of sample C in figure 1 in terms of the concentration of the precursor solutions, as they were derived from solutions with the same pH value under the same flow rate of 3 ml·min^{-1}. Sample D from the system of pH 8 consisted of rectangular sheet-like particles, exhibiting a marked contrast to the morphology of samples B and C. The nano-sheets were 300 nm or longer in length and ~70 nm wide. Sample F showed heavy agglomeration of the primary particles as did sample C, yet the primary particles of sample F looked needle-like, while

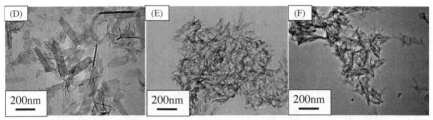

Figure 3. TEM micrographs of the samples, prepared from the concentrated solution system (50 mM $Ca(NO_3)_2 \cdot 4H_2O$ and 30 mM $(NH_4)_2HPO_4$; Flow rate 3 ml·min^{-1}; 40°C. (D) pH8, (E) pH9, (F) pH10

those of sample C were spherical. Similar agglomeration of needle-like particles was observed for sample E. All X-ray diffractions of samples D, E, and F shown in figure 4 were assigned to HA, without any by products, as were those from the dilute solution system. Moreover, all samples gave practically the same XRD profile, indicating that crystallinity was independent of the pH of the concentrated solution system though the crystallite shape was dependent on it.

As no XRD signals were detectable for sample A, a selected area electron diffraction (SAD) was taken for sample A and is shown in figure 5. Very faint but distinct diffraction rings were confirmed. Thus, the particles of the sample were tinged with crystallinity: it is not yet certain if they involved any amorphous regions or if they were fully crystalline. The morphology of sample D changed by having the lattice structure of standard hydroxyapatite. Note that a particle shape can dramatically change by having the crystal structure and stoichiometric composition of HA when the synthesis condition is changed.

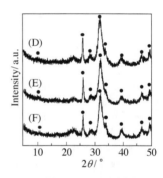

Figure 4. XRD patterns of the samples, prepared from the concentrated system. (D)

(iii) Mechanisms of precipitation and particle shapes

The deposition mechanisms for single nano-particles like sample A and rectangular nano-sheets like sample D were different. Even in dilute solution systems, many nuclei are formed under pH 10. Yet, the lowest flow rate not only carries a smaller amount of calcium and phosphate ions which the nuclei capture to grow, but also provides very little probability that a few

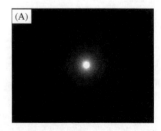

Figure 5. Photograph of selected area electron diffraction (SAD) pattern for sample A.

nuclei collide with each other to lead to agglomeration. Greater flow rates, then, provide larger probability of such collision, which results in agglomerated particle shapes, as observed in figure 1 (C).

In concentrated solution systems with high pH, 9 and 10, particles with a needle-like shape were obtained, since such a morphology of HA crystallites is common to those from large-scale wet-chemical synthesis routes. The high flow rate (3 ml·min^{-1}) and high pH should stimulate a greater rate of nucleation and growth of HA. Thus, the precipitation reaction modes in the present micro-reactor are practically the same as in common wet-chemical synthesis. In contrast, the particles of sample D were precipitated at the lowest level of the pH range where HA is stable. In this case, the HA nuclei should be formed with only a slight margin of thermo-chemical stability as an equilibrium of precipitation-dissolution is established. That is, it may take a longer time for them to grow, or the calcium and phosphate ions to be settled might choose their favorite sites. It is unfortunate that the orientation of the sheet, or which hkl plane the HA moiety is exposed to, is not known now. Detailed electron diffraction studies are under progress. Yet, the sheet-like crystallites provide large flat surfaces, which may adsorb many kinds of physiological substances.

CONCLUSIONS

It was proposed that a single nano-particle and nano-sheet of HA could be prepared by using a micro reactor applying the micro chemical process. It was found that the morphology of the particles was dramatically changed by changing the pH, flow rate, and ion concentration. The dilute solution system yielded spherical HA particles, whose average particle size increased from about 3 nm to 40 nm, when the pH was kept at 10. The concentrated solution system yielded rectangular, plate-like HA particles, 300 nm long and 50 nm wide, when the pH was kept at 8. These particles, with characteristic shapes, are expected to be applied not only as implant materials but also as DDS carriers.

ACKNOWLEDGEMENT

Financial support by the Electric Technology Research Foundation of Chugoku is gratefully acknowledged.

REFERENCES

[1] D. A. Puleo, L. A. Holleran, R. H. Doremus, and R. Bizios, Osteoblast responses to orthopaedic implant materials in vitro, *J. Biomed. Mater. Res.*, 25, 711-723 (1991).

[2] R. Garcia, R. H. Doremus, Electron microscopy of the bone-implant interface from a human dental implant, *J. Mater. Sci. Mater. Med.*, 3, 154-156 (1992).

[3] A. Tiselius, S. Hjerten, and O. Levin, Protein chromatography on calcium phosphate columns, *Arch. Biochem. and Biophys.*, 65, 132-155 (1956).

[4] T. Kawasaki, K. Ikeda, S. Takahashi, and Y. Kuboki, Further study of hydroxyapatite high-performance liquid chromatography using both proteins and nucleic acids, and a new technique to increase chromatographic efficiency, *Eur. J. Biochem.*, 155, 249-257 (1986).

[5] T. Kawasaki, W. Kobayashi, K. Ikeda, S. Takahashi, and H. Monma, High-performance liquid chromatography using spherical aggregates of hydroxyapatite micro-crystals as adsorbent, *Eur. J. Biochem.*, **157**, 291-295 (1986).

[6] S. Takashima, S. Hayakawa, C. Ohtsuki, and A. Osaka, Adsorption of proteins by calcium phosphate with varied Ca to P ratios, *Bioceramics*, **9**, 217-220 (1997).

[7] S. Hayakawa, U. Kusudo, S. Takemoto, K. Tsuru, and A. Osaka, Hydroxy-carbonate apatite, blood compatibility and adsorption of specific pathogenic proteins, *Bioceramics Mater. and Appl. IV*, **147**, 111-119 (2003).

[8] S. Takemoto, U. Kusudo, K. Tsuru, S. Hayakawa, A. Osaka, and S. Takashima, Selective protein adsorption and blood compatibility of hydroxy-carbonate apatites, *J. Biomed. Mater. Res.*, **69A**, 544-551 (2004).

[9] C.A. Zittle, Adsorption studies of enzymes, and other proteins, *Ad. in Enzymol*, **14**, 319-374 (1951).

[10] E. Fujii, M. Ohkubo, K. Tsuru, S. Hayakawa, and A. Osaka, Selective protein adsorption property and characterization of nano-crystalline zinc-containing hydroxyapatite, *Acta Biomater.*, **2**, 69-74 (2006).

[11] S. Hayakawa, A. Osaka, K. Tsuru, K. Ando, E. Fujii, K. Kawabata, C. Bonhomme, and F. Babboneau, Synthesis and Characterization of Mg-Containing Nano-Apatite, *Key Eng. Mater.*, **361**, 47-50 (2006).

[12] E. Fujii, K. Kawabata, H. Yoshimatsu, K. Tsuru, S. Hayakawa, and A. Osaka, Synthesis and Structural Characterization of Silica-Hybridized Hydroxyapatite with Gas Adsorption Capability, *J. Ceram. Soc. Japan*, **9**, 769-773 (2006).

[13] Powder Diffraction File, Card No.9-423, Joint Committee on Powder Diffraction Standards.

SYNTHESIS, MECHANICAL AND BONE CELL MATERIALS INTERACTION STUDIES ON SRO AND MGO DOPED RESORBABLE TRICALCIUM PHOSPHATE FOR BONE TISSUE ENGINEERING

Shashwat Banerjee, Amit Bandyopadhyay and Susmita Bose*
W. M. Keck Biomedical Materials Research Laboratory, School of Mechanical and Materials Engineering, Washington State University
Pullman, WA 99164, USA.
*sbose@wsu.edu

ABSTRACT
Calcium phosphates are of significant interest due to their chemical similarity with the inorganic component of natural bone. We have used different dopants, such as Sr^{2+}, Mg^{2+}, Si^{2+}, and Zn^{2+} to study the effects of dopant chemistry on resorbability and strength loss behavior of TCP. Dopants were introduced in TCP both by chemical synthesis and physical mixing methods with an amount of <5wt% of TCP amount. For example, addition of binary dopant composition, such as Mg^{2+} and Sr^{2+}, improved compression strength of TCP from 56.71 MPa to 107.34 MPa. Human osteoblast cell culture was used to determine the influence of dopants on cell-materials interactions, e.g. cell proliferation by MTT. Osteoblast culture studies indicated that the presence of dopant favored cell growth. Our study on both *in vitro* and *in vivo* resorption behavior of these doped TCP show as a promising material for bone tissue engineering.

INTRODUCTION

Tricalcium phosphate (TCP) is one of the most commonly used materials in biomedical applications due to the close chemical similarity to the inorganic component of bone and teeth mineral. Due to its high biocompatibility, unique bioresorbable features and excellent osseoconductive properties TCP has found wide spread use as bone cement and implant materials [1-3]. Furthermore, TCP has the ability to degrade under physiological conditions of the body to allow bone regeneration. It can be completely replaced by the newly grown tissue thus avoiding second surgery. Although there are many reasons for the use of TCP, there are still critical issues concerning the use of these ceramics. Primarily, low mechanical properties limit its application in load bearing applications.

Trace amount of Na, Mg, Zn, Fe, Cu, Pb, Sr, and Si are found in bones and teeth. These trace elements play an important role in overall performance of human bone. Under this perspective, substitution of trace elements into the calcium phosphate structure has been the subject of widespread investigation nowadays because of their impending role in the biological process during implantation studies. In recent years, substitution of magnesium (Mg) in the calcium phosphate has received much attention due to its impending role in the qualitative changes in bone matrix, bone metabolism, promoting catalytic reactions, and controlling biological functions [4, 5].

Strontium has a beneficial effect on bone. In recent years, strontium has been gradually recognized during the research of treatment for osteoporosis. It enhances the replication of preosteoblastic cells, and simulates bone formation in calvarial cultures in vitro [6]. Furthermore, it has been demonstrated that strontium ranelate decreased bone resorption in vivo [7, 8]. As a kind of bone-seeking trace elements, strontium has various effects on bone metabolism depending on the dose used. At low dose levels, stable strontium is of great benefit to bone formation as mentioned above.

In our previous studies, the individual effects of Mg or Sr substitution on morphology, dissolution properties and cellular biocompatibility of calcium phosphate have been documented [4, 9-11]. However, very few studies on the effect of Mg or Sr substitution in β-tricalcium phosphate has been reported. To the best of our knowledge no study has been reported on the simultaneous doping of Mg and Sr in β-tricalcium phosphate. The purpose of this study was to determine the effects of simultaneously incorporation of Mg and Sr on the physico-chemical properties of β-tricalcium phosphate. Results obtained from this study may provide a better understanding for designing doped β-tricalcium phosphate with desired physico-chemical and biological properties as bone graft material.

MATERIALS AND METHODS

Calcium nitrate ($Ca(NO_3)_2 \cdot 4H_2O$), magnesium nitrate ($Mg(NO_3)_2 \cdot 6H_2O$, and strontium oxide ($Sr(NO_3)_2 \cdot 2H_2O$) were purchased from J. T. Baker (NJ). Phosphoric was procured from Fisher Scientific (NJ). Ammonium hydroxide (NH_4OH) was purchased from J. T. Baker, NJ. Hydrazinium hydroxide was obtained from Merck (Germany). The water used throughout this work was distilled water. All other chemicals were of analytical grade and used without further purification.

SYNTHESIS OF TRICALCIUM PHOSPHATE (TCP)

Four compositions of TCP were synthesized in this study: TCP, TCP doped with 1 wt% MgO (1% Mg-TCP), TCP doped with 4 wt% SrO (4% Sr-TCP) and TCP doped with the combination of 1 wt% MgO and 4 wt% SrO (1% Mg-4% Sr-TCP). For Mg and Sr doped TCP samples, (Ca+Mg)/P, (Ca+Sr)/P and (Ca+Mg+Sr)/P molar ratio are kept at 1.5. TCP powder was synthesized by a wet coprecipitation method according to our previous work [9]. 100 mL of 1.0 M Ca^{2+} aqueous solution was mixed with 0.667 moles of phosphoric acid to maintain Ca^{2+} to PO_4^{3-} ratio 1.5 to 1, similar to TCP. The pH of the precursor solution was between 2 and 3. Ammonium hydroxide (NH_4OH) was added drop wise with continuous stirring on a magnetic stirrer to gradually increase pH of the solution. Final pH of the solution was maintained at 7.0. Resulting precipitate was filtered and dried at 120 °C for 12 h. After drying, precipitate was grinded into fine powder and placed in the furnace for calcinations at 650°C for 2 h. 1% Mg-TCP, 4% Sr-TCP and 1% Mg-4% Sr-TCP powders were also prepared by adding required amount of magnesium nitrate and strontium nitrate in the solution, and the same precipitation process was followed.

CHARACTERIZATION

The powders obtained were pressed by uniaxial pressing at a pressure of 145 MPa and by CIP (Cold isostatic pressed) at 48 ksi, and then sintered at 1250 °C for 2 h, respectively. The surface morphologies and microstructures of sintered samples were observed using a scanning electron microscope (SEM, Hitachi S-570, Japan). Phase analyses were performed by an X-ray diffractometer (Philips PW 3040/00 X'pert MPD). Fourier transform infrared (FTIR) spectra were recorded on a ThermoFisher FTIR spectrometer (Nicolet 6700, ThermoFisher, Madison, WI). The relative density was determined by the Archimedian's method. The compressive strength was measured using a mechanical test machine (Instron) with a loading rate of 0.5 mm/min.

MTT ASSAY

All samples were sterilized by autoclaving at 121 °C for 30 min. An established human osteoblast cell line hFOB 1.19 (ATCC, Manassas, VA) was used in this study. Cells were seeded onto the samples placed in 24-well plates. The base medium for this cell line is a 1:1 mixture of Ham's F12 Medium and Dulbecco's Modified Eagle's Medium (DMEM/F12, Sigma, St. Louis, MO), with 2.5 mM L-glutamine (without phenol red). The medium was supplemented with 10% fetal bovine serum (HyClone, Logan, UT) and 0.3 mg/ml G418 (Sigma, St. Louis, MO). Cultures were maintained at 34°C under an atmosphere of 5% CO_2. The medium was changed every 2-3 days for the duration of the experiment.

MTT (3-(4,5-dimethylthiazol-2-yl)-2,5-diphenyl tetrazolium bromide) assay was used to evaluate cell proliferation. The MTT (Sigma, St. Louis, MO) solution of 5 mg/ml was prepared by dissolving MTT in PBS, and was filter sterilized using a filter paper of 0.2 micron pore. The MTT was diluted (50 μl into 450 μl) in DMEM/F12 medium. 500 μl diluted MTT solution was then added to each sample in 24-well plates. After 2 h of incubation, 500 μl of solubilization solution made up of 10% Triton X-100, 0.1N HCl and isopropanol were added to dissolve the formazan crystals. 100 μl of the resulting supernatant was transferred into a 96-well plate, and read by a plate reader at 570 nm.

IN VITRO MINERALIZATION STUDY BY SIMULATED BODY FLUIDS (SBF) IMMERSION

Influence of Mg and Sr doping on in vitro mineralization behavior of tricalcium phosphate samples were evaluated by immersion in simulated body fluids (SBF) which has a similar ionic concentration to human blood plasma [12]. The dissolution and mineralization behavior were evaluated by immersion of samples made with synthesized powders in simulated body fluid (SBF) having concentration: 2.5 mM of Ca^{2+}, 1.5 mM of Mg^{2+}, 142.0 mM of Na^+, 5.0 mM of K^+, 148.5 mM of Cl^-, 4.2 mM of HCO_3^-, 1.0 mM of HPO_4^{2-}, 0.5 mM of SO_4^{2-}. The SBF solution was buffered at pH 7.40 with 50 mM tri (hydroxymethyl) aminomethane $((CH_2OH)_3CNH_2)$ and approximately 45 mM hydrochloric acid (HCl) at 37 °C. SBF solution was changed twice every week with freshly made solutions. Samples were taken out of SBF at a definite time interval, gently rinsed twice in 100 ml distilled water, placed on a porous alumina plate, and dried in oven at 60 °C for 72 h. The masses of the samples were measured after immersion for 7 days. Surface microstructures of the samples were observed with FE-SEM.

RESULTS AND DISCUSSIONS

Phase identification and evolution

The XRD patterns for all the powders calcined at 650°C and sintered at 1250°C are presented in figure 1. Samples sintered at 1250°C showed improvement in crystallinity by the increase in the resolution of the peaks when compared to the powders calcined at 650°C. At 1250°C, all the powders indicated formation of β-TCP as the major phase (JCPDS No 9-169). However, the peaks pertaining to the 1% Mg doped β -TCP and 1% Mg-4% Sr doped TCP did not show exact match with those of JCPDS No. 09-169, due to the shift in the 2θ and d-spacing values of maximum intensity plane (0210). This may be because of the ionic radius of Mg and Sr. The ionic radius of Sr^{2+} (1.13A°) was slightly larger than that of Ca^{2+} (0.99A°) but for Mg the ionic radius was smaller (0.66 Å) [12]. Because of the smaller ionic radius, Mg–O distances was shorter compared to that of Ca–O bond. Stronger Mg–O interactions than Ca–O influences lattice constant values leading to lattice contraction and shift in maximum intensity plane (0210) [14].

FTIR spectra of β-TCP samples are shown in figure 2. All the samples showed vibrational modes characteristic of PO_4^{3-}. The bands at 559 cm^{-1} and 613 cm^{-1} were assigned to the O–P–O bending mode, 1070 and 1020 cm^{-1} to asymmetric stretching mode and 932 cm^{-1} to symmetric stretching vibration. The new peaks observed in the spectra of Mg doped TCP and Sr-doped TCP at 504 cm^{-1} and 561 is due to Mg–O and Sr–O stretching frequency. The P-O bands at 1000-1100 cm^{-1} became sharper and narrower with Mg doping in the TCP. This indicates that the crystallinity has improved and the crystallite size became larger. The frequency of the Ca–O bond decreased with Mg and Sr doping in TCP. These changes in frequencies are consistent with the Mayer bond orders, and suggest that the Ca–O bond weakens with the addition of Mg, consistent with the trend in the geometrical structure [14].

The relative density of samples sintered at 1250°C was evaluated and are shown in figure 3. The theoretical density (TD) of undoped β-TCP was found to be 81.4% but on doping with 1% Mg the density increased to 87.2% which showed that Mg doping helped in densification of TCP. TCP doped with dual dopant, Mg and Sr showed the maximum of 95.66% of TD. In usual cases, pure β-TCP without any additions in its lattice is not stable beyond 1120°C and tends to transform into α-polymorph of TCP which prevents densification due to volume expansion during phase transformation. The higher density of Mg doped TCP samples can be due to the influence of Mg on the phase stability of TCP. Mg is found to stabilize the β-TCP phase at higher temperature. The improved thermal stability of the β-TCP phase with substituted Mg resulted in increase of density of the Mg doped TCP samples.

The microstructure of the synthesized powders was examined by SEM. The SEM micrographs of synthesized powders sintered at 1250°C are shown in figure 4. Undoped TCP, and 1% Sr doped TCP showed porous microstructure as can be seen in figure 3a and 3c. Addition of Mg increased the density of sintered TCP sample as shown in figure 3b. The micrograph (figure 3d) clearly revealed that 1% Mg-4% Sr-TCP sintered well and an almost fully dense microstructure was obtained.

Compressive strengths of all samples sintered at 1250°C are presented in figure 5. Undoped TCP showed low strength. Addition of Mg or Sr improved the compressive strength of TCP negligibly. Highest compressive strength was obtained for 1% Mg-4% Sr-TCP. The results clearly showed that the individual dopant effects on compressive strength may be significantly different compared with their combined effects. Mg or Sr additions did not significantly improve compressive strength of TCP, however, their combined addition increased the compressive strength significantly. The primary reason for strength variations can be due to grain boundary strengthening and change in densification.

CELL CULTURE

In vitro cytotoxicity behavior of doped β-TCP samples were evaluated using human fetal osteoblast cells (hfOB). The MTT assay was used to determine hfOB cell proliferation on β-TCP and doped β-TCP. Figure 6 shows the comparison of cell densities on TCP, 1%Mg-TCP, 4%Sr-TCP and 1%Mg-4%Sr-TCP after 3, 5 and 11 days of culture. As evident from the figure, cell numbers increased with increasing culture time. After 3 days of culture, all doped samples showed higher cell density compared to simple TCP. The cell proliferation was more on Mg and Sr doped TCP as compared to all other samples as was obvious for the 11 day sample. This result suggests that the presence of Mg and Sr enhances cellular attachment and growth.

APATITE FORMING ABILITY IN SBF

In vitro mineralization study of TCP and doped TCP samples were performed by immersion in simulated body fluid solution (SBF) for 7 days. Figure 7(a, c) shows significant growth of apatite layer on the surface of undoped β-TCP and 4% Sr-TCP. Figure (b, d) shows microstructures of 1% Mg-TCP and 1% Mg-4% Sr-TCP. No apatite formation was visible on the surface of Mg-TCP and very little apatite formation was observed on 1% Mg-4% Sr-TCP. This can be explained on the basis of density and mechanical strength of the samples. Low density and mechanical strength are important factors for fast dissolution [15, 16]. Fast dissolution lead to a SBF solution with higher Ca^{2+} and PO_4^{2-} along with other anionic groups i.e., HPO_4^{2-}, HCO_3^-, OH^- triggering accelerated nucleation and growth of apatite. TCP and 4% Sr-TCP showed low density and mechanical strength as compared to 1% Mg-TCP and 1% Mg-4% Sr-TCP. Hence bone like apatite layer was observed on the surface of TCP and 4% Sr-TCP.

CONCLUSIONS

TCP and doped TCP were successfully synthesized with 1% Mg and 4 % Sr dual dopants using aqueous precipitation method. Addition of dual dopants Mg and Sr improved compression strength of TCP from 56.71 MPa to 107.34 MPa. The synthesized powders characterized by XRD indicated that stronger Mg–O interactions influenced the lattice constant values leading to lattice contraction in dual doped TCP. Human osteoblast cell culture was used to determine the influence of dopants on cell-materials indicated that the presence of dopant favored cell growth. The study reveals the potential of dual doped TCP with Mg and Sr doped as a promising material for the bone tissue engineering.

ACKNOWLEDGEMENTS
Authors like to acknowledge the financial support from the National Institutes of Health (Grant # NIH-R01-EB-007351).

REFERENCES
1. R. D. Gaasbeek, H. G. Toonen, R. J. van Heerwaarden, P. Buma, Mechanism of bone incorporation of β-TCP bone substitute in open wedge tibial osteotomy in patients, *Biomaterials*, **26(33)**, 6713–6719 (2005).
2. N. Kondo, A. Ogose, K. Tokunaga, T. Ito, K. Arai, N. Kudo, H. Inoue, H. Irie, N. Endo, Bone formation and resorption of highly purified β-tricalcium phosphate in the rat femoral condyle, *Biomaterials*, **26(28)**, 5600–5608 (2005).
3. R. Tang, W. Wu, M. Haas, G. H. Nancollas, Kinetics of Dissolution of b- Tricalcium Phosphate, *Langmuir*, **17[11]**, 3480–3485 (2001).
4. A. Bandyopadhyay, S. Bernard, W. Xue and S. Bose, "**Feature Article**: Calcium Phosphate Based Resorbable Ceramics: Influence of MgO, ZnO and SiO2 Dopants," *Journal of the American Ceramic Society*, **89** [9], pp. 2675-88 (2006).
5. Y. Yamasaki, Y. Yoshida, M. Okazaki, A. Shimazu, T. Uchida, T. Kubo, Y. Akagawa, Y. Hamada, J. Takahashi, and N. Matsura, Synthesis of Functionally Graded MgCO3 Apatite Accelerating Osteoblast Adhesion, *J. Biomed. Mater. Res.*, **62(1)**, 99–105 (2002).
6. E. Canalis, M. Hott, P. Deloffre, Y. Tsouderos, P. J. Marie, The divalent strontium salt S12911 enhances bone cell replication and bone formation in vitro. *Bone*, **18(6)**, 517–523 (1996).
7. J. Buehler, P. Chappuis, J. L. Saffar, Y. Tsouderos, A. Vignery, Strontium ranelate inhibits bone resorption while maintaining bone formation in alveolar bone in monkeys, *Bone*, **29(2)**,176–179 (2001).
8. M. Hott, P. Deloffre, Y. Tsouderos, P. J. Marie, S12911-2 reduces bone loss induced by short-term immobilization in rats, *Bone*, **33**, 112–123 (2003).
9. Weichang Xue, Kelli Dahlquist, Ashis Banerjee, Amit Bandyopadhyay, Susmita Bose, Synthesis and characterization of tricalcium phosphate with Zn and Mg based dopants, *J Mater Sci: Mater Med*, **19**, 2669–2677 (2008).
10. W. Xue, J. L. Moore, H. L. Hosick, S. Bose, A. Bandyopadhyay, W. W. Lu, Kenneth, M. C. Cheung, Keith D. K. Luk, Osteoprecursor cell response to strontium-containing hydroxyapatite ceramics, *Journal of Biomedical Materials Research Part A*, **79A(4)**, 804-814 (2006).
11. W. Xue, H. L. Hosick, A. Bandyopadhyay, S. Bose, C. Ding, K DK Luk, K MC Cheung, W W Lu, "Preparation and cell-materials interactions of plasma sprayed strontium-containing hydroxyapatite coating," *Surface and Coatings Technology, Volume* **201** [8], pp. 4685-93 (2007).
12. K. Qiu, X. J. Zhao, C. X. Wan, C. S. Zhao, Y. W. Chen, *Biomaterials*, **27**, 1277–1286, (2006).
13. Kokubo T, Takadama H. How useful is SBF in predicting in vivo bone bioactivity? *Biomaterials*, **27**, 2907-2915 (2006).

14. X. Yin, L. Calderin, M. J. Stott, M. Sayer, Density functional study of structural, electronic and vibrational properties of Mg- and Zn-doped tricalcium phosphate biomaterials, *Biomaterials*, **23**, 4155–4163 (2002).
15. W. R Lacefield. *An introduction to bioceramics*. World Scientific, London: L. L. Hench, J. Wilson, 76-89 (1993).
16. S. Mann, Molecular recognition in biomineralization, *Nature*, **332**, 119–124 (1988).

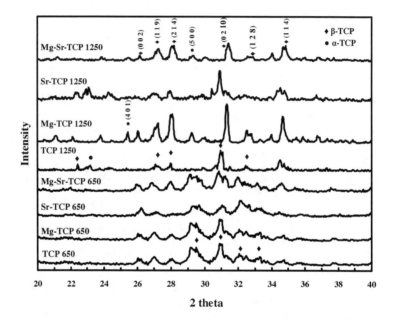

Figure 1. XRD spectra of TCP, 1% Mg-TCP, 4% Sr-TCP and 1% Mg-4% Sr-TCP samples calcined at 650°C and sintered at 1250°C.

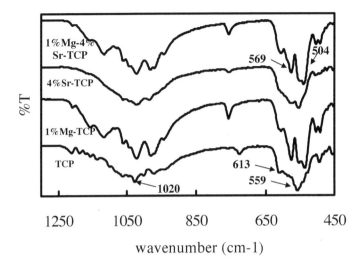

Figure 2. FTIR spectra of TCP, 1% Mg-TCP, 4% Sr-TCP and 1% Mg-4% Sr-TCP samples sintered at 1250°C.

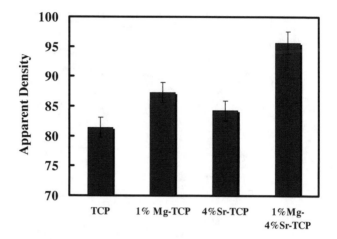

Figure 3. Apparent density of TCP, 1% Mg-TCP, 4% Sr-TCP and 1% Mg-4% Sr-TCP samples sintered at 1250°C.

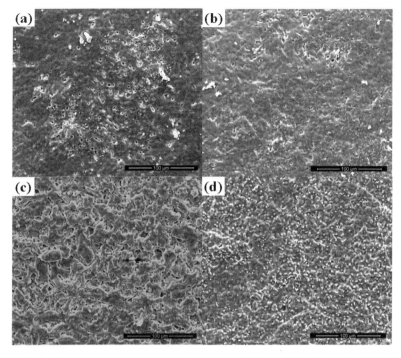

Figure 4. SEM surface morphologies of samples sintered at 1250°C. (a) TCP, (b) 1% Mg-TCP, (c) 4% Sr-TCP and (d) 1% Mg-4% Sr-TCP

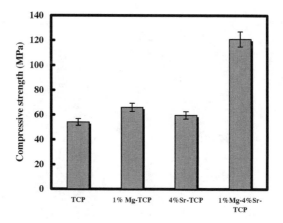

Figure 5. Compressive strength of TCP, 1% Mg-TCP, 4% Sr-TCP and 1% Mg-4% Sr-TCP samples sintered at 1250°C.

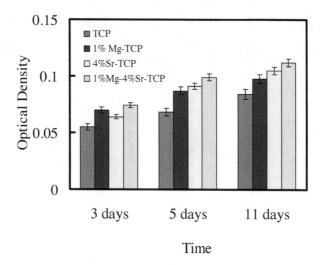

Figure 6. Optical density measurement illustrating human fetal osteoblast cells (hFOB) proliferation on TCP, 1% Mg-TCP, 4% Sr-TCP and 1% Mg-4% Sr-TCP for 3, 5 and 11 days of culture time.

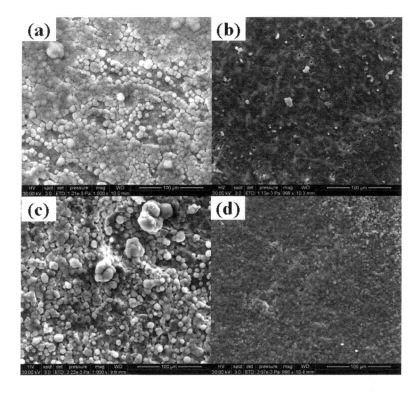

Figure 7: SEM micrographs of the precipitation of apatite layers after immersion in SBF for 7 days on (a) TCP, (b) 1% Mg-TCP (c) 4% Sr-TCP and (d) 1% Mg-4% Sr-TCP.

Al$_2$O$_3$ and TiO$_2$

CELL PROLIFERATION ON NANO-STRUCTURED TITANIUM OXIDE LAYER PREPARED ON TITANIUM SUBSTRATES

Satoshi Hayakawa, Koji Kawanishi, Yuki Shirosaki, Akiyoshi Osaka
Biomaterials Laboratory, Graduate School of Natural Science and Technology, Okayama University
Research Center for Biomedical Engineering, Okayama University
3-1-1, Tsushima, Kita-ku, Okayama, 700-8530, Japan.

ABSTRACT
    Effects of nanometer-scale surface roughness of titanium oxide layer on the adhesion and proliferation of osteoblast-like cell were examined. The nano-structured titanium oxide layer was prepared on titanium substrates by chemical treatment and heat-treatment. The anatase or rutile phase of titanium oxide was formed on the titanium substrates. The surface roughness of titanium oxide layer depended on the chemical treatment time or the heat-treatment time. The average roughness (Sa) was controlled in the range from 10 to 70 nm. The cell metabolic activity increased on the substrate with limited surface roughness. It is found that cell proliferation is affected by nanometer-scale surface roughness of titanium oxide layer.

INTRODUCTION
    Titanium and titanium-based alloys have widely been applied to orthopaedic and dental implant because of their good mechanical strength and corrosion resistance. Many researchers have reported the method for providing the material surface with the apatite-forming ability, because the deposition of bone-like apatite on the surface of materials is considered to be prerequisite when the implant materials can bond to bone tissue.[1] Among them, Wang et al. [2,3] reported that crystalline (anatase) TiO_2 layer prepared through chemical and subsequent heat-treatment of titanium substrates had a superior in vitro apatite-forming ability. This method was denoted as a CHT method.[4] Recently, Sugino et al.[5-7] reported the effect of special gap on heterogeneous nucleation of bone-like hydroxyapatite (BHAp) on the surface of thermally oxidized titanium substrates, and revealed the importance of designing the surface morphology of titanium implants, followed by oxidization, to induce osteoconductivity, and use of spatial design of the titanium implant.
    The cell adhesion is one of the most critical initial events occurring during the interactions of cells with implants and has further influence on the proliferation and differentiation of bone cells before bone tissue formation.[8] In this study, we pay attention to two methods such as the CHT method[4] and the thermal oxidation process of titanium substrates and produced reproducible nanometer-scale surface roughness. The nano-structured titanium oxide layer was prepared on mirror-polished titanium substrates by the CHT method or thermal oxidation process, respectively, in order to examine the effect of nanometer-scale surface roughness of titanium oxide layer on the initial adhesion and

proliferation of osteoblast-like cell; MC3T3-E1. The cell metabolic activity was correlated to the various surface features such as crystal phase and topographic parameter (average roughness, Sa).

EXPERIMENTAL PROCEDURE

Commercial available mirror-polished titanium discs (15mmφ, pure Ti, Grade II) were washed three times in acetone for 5 min in an ultrasonic cleaner (coded as NT). Anatase-type TiO_2 layer was prepared on pure titanium discs by chemical treatment with 3% H_2O_2 solution at 80°C for various periods ranging from 0.5 up to 4h and subsequent heat-treatment at 400°C for 1h in an electric furnace in air (coded as CHTn: n indicates chemical treatment time (hour) ; n= 0.5, 1, 2, 3 or 4). Rutile-type TiO_2 layer was prepared on pure titanium discs by thermal oxidation process as follows: NT was heat-treated at 600°C for various periods ranging from 1 hour up to 24 hours in an electric furnace in air. Obtained samples were coded as HTn (n indicates heat-treatment time (hour) ; n = 1, 3, 6, 12 or 24).

Surface characterization

Surface crystal structure was examined by a thin-film X-ray diffractometry (TF-XRD, RINT2500, Rigaku Co., Tokyo, Japan; CuKα, 40 kV, 200 mA, a scanning rate of 0.02° *per* sec for 2θ scan), where the angle of incident X-ray beam, θ was fixed at 1.0°.

Three-dimensional images of the sample were taken by using an atomic force microscope (AFM, Nanopics 2100, SII NanoTechnology Inc., Osaka, Japan). A software SPIP[TM] (Image Metrology A/C, Denmark) was used to analyze topography of titanium substrates in terms of parameters such as average roughness (Sa), roughness root mean square (Sq), the maximum height between the peak (highest) and valley (lowest) (Sz) and the increment of the interfacial surfaces area relative to the area of the projected (flat) x, y plane (Sdr).

Biological assessment

MC3T3-E1 murine calvaria osteoblast-like cells (Riken Cell Bank, Ibaraki, Japan) were cultured in alpha modification of eagle's medium (α-MEM), supplemented with 10 vol% fetal bovine serum (FBS), 1 vol% penicillin-streptomycin (10,000 U/ml), 1 vol% Fungizon (250 µg/ml), 1vol% ascorbic acid (5 mg/ml). At confluences, cells were detached from the culture dishes by Actinase/EDTA followed centrifugation (2200 rpm, 5min). They were re-suspended in the culture medium. Before the cell seeding, the samples were autoclaved at 121°C for 20 min and dried for 90 min. Cells were seeded at density of 2×10^4 cells/well on CHTn samples, while cells were seeded at density of 1×10^4 cells/well on HTn samples. Thermanox[®] was used as a positive control. Cells were kept in culture at 37°C and 5% CO_2 for various periods. alamarBlue[®] assay was used for the evaluation of cellular viability or metabolic activity.[9] At days 1, 2, 3 and 5, a mixture of alamarBlue[®] (in an amount equal to 10% of the total culture volume) was aseptically added to the samples. Samples were incubated with alamarBlue® for an additional 3h. alamarBlue® is chemically reduced by the metabolic activity of growing cells, which causes the fluorometric-colorimetric REDOX indicator to change from an oxidized, nonfluorescent blue form to a reduced, fluorescent red form. The fluorescence intensity was measured by using a Multi-Microplate Reader (MTP-800Lab, CORONA ELECTRIC Co., Ltd.) with excitation at 550 nm and emission at 615 nm according to the manufacturer's direction. The obtained values were compared with those of the positive control.

RESULTS & DISCUSSION

Fig.1 showed the TF-XRD patterns of the titanium samples. CHTn samples showed the 101 diffraction of anatase phase (PDF #21-1272) at 26° in 2θ, while HTn samples showed diffractions of rutile phase (PDF #21-1276) at ca. 27.4°, 36.1° and 39.2° in 2θ, together with diffractions of titanium (α-Ti; PDF #44-1294) at ca. 35.1° and 38.4° in 2θ. The intensity of the 101 diffraction of anatase increased with the chemical treatment time. Similarly, the intensity of the 110 diffraction of rutile increased with the heat-treatment time.

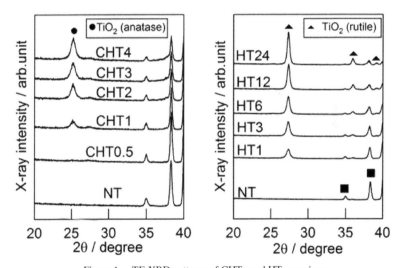

Figure 1. TF-XRD patterns of CHTn and HTn specimens.

Fig. 2 shows representative AFM images of the surface morphology of the titanium samples. Although the surface of NT or Thermanox® was very smooth, HTn surfaces showed polygonal regions corresponding to the grains of titanium metal (α-Ti) and were full-covered with nano-topographic rod arrays of TiO_2 (rutile) crystals. The increment of the interfacial surfaces area relative to the area of the projected (flat) x, y plane (Sdr) increased with increasing the chemical treatment time or heat-treatment time.

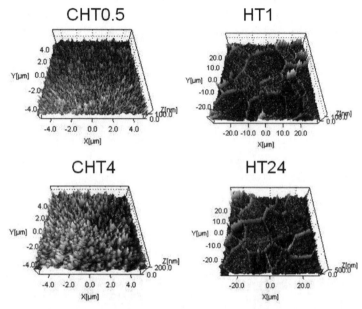

Figure 2.   Representative AFM images of CHTn and HTn specimens.

As shown in Fig.3, the average roughness (Sa) and the maximum height between the peak and valley (Sz) of CHTn gradually increased with increasing chemical treatment time. On the other hand, Sa and Sz values of HTn monotonically increased with increasing heat-treatment time. Judging from the Sz (nm) and AFM images (Fig.2), the TiO_2 (rutile) crystals especially grew up well at the grain boundary.

Fig.4 showed the cell metabolic activity of MC3T3-E1 cell on CHTn and HTn specimens. The cell metabolic activity of CHTn specimens until 5 day was lower than the positive control (Thermanox®). The cell metabolic activity was almost same among HTn specimens until 1 day. The cell proliferation was improved on HT6 after 2 days and on HT12 after 3 days compared with the positive control (Thermanox®).

In vitro and in vivo studies clarified that osteoblasts attach and spread more rapidly on rough surfaces than smooth ones and the adhesion of osteoblasts is enhanced on titanium surfaces with 1-10 μm-size roughness.[10-14]   On the other hand, Webster et al.[15] reported that the nanometer-scale surface roughness ranging from 17 to 32 nm is a crucial that determines the cytocompatibility of nanoceramics in terms of cell adhesion, and the cellular responses to nanophase ceramics is independent of surface chemistry and the material crystalline phase but is dependent on the surface topography.   The fact that HT6, the average roughness of 36.6 ± 7.0 nm, showed a high MC3T3-E1 cell metabolic activity indicated that the nanometer-scale surface roughness ranging from 37-52 nm due to characteristic topography of titanium oxide layer is favorable for the MC3T3-E1 cell metabolic activity.

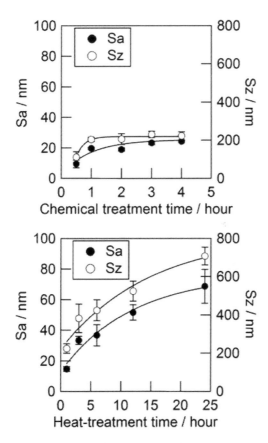

Figure 3.   Average roughness (Sa) and the maximum height between the peak and valley (Sz) of CHTn and HTn specimens.

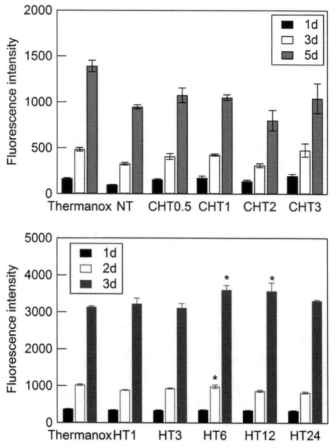

Figure 4.   Cell metabolic activity of CHTn and HTn specimens.
Asterisks (*) indicate a significant difference from Thermanox® (p ≦ 0.01))

CONCLUSION

We prepared the nano-structured titanium oxide layer on mirror polished titanium substrates by chemical treatment and heat-treatment and examine the effect of nanometer-scale surface roughness on the adhesion and proliferation of osteoblast-like cell; MC3T3-E1.   The anatase or rutile phase of titanium oxide was formed on the titanium substrates.   The average roughness (Sa) was controlled in the range from 15 to 70 nm by the chemical and heat-treatment time. The cell metabolic activity increased on the substrate heated for 6 or 12 hours, indicating that the cell adhesion and proliferation is affected by nanometer-scale surface roughness ranging from 37-52 nm.

ACKNOWLEDGEMENT
    This study was supported by a Grant-in-Aid for Science Research from the Japan Society for the Promotion of Science (19360301) and the grant of Nippon Sheet Glass Foundation for Materials Science and Engineering.

REFERENCES
[1]T. Kokubo, H. Kushitani, S. Sakka, and T. Yamamuro, Solutions Able to Reproduce *in vivo*-Surface-Structure Changes in Bioactive Glass-Ceramic A-W, *J. Biomed. Mater. Res.*, **24**, 721–734 (1990).

[2]X-X. Wang, S. Hayakawa, K. Tsuru, and A. Osaka, Improvement of bioactivity of $H_2O_2/TaCl_5$-treated titanium after subsequent heat treatments, *J. Biomed. Mater. Res.*, **52**, 171-176 (2000).

[3]X-X. Wang, S. Hayakawa, K. Tsuru, and A. Osaka, Bioactive Titania Gel Layers Formed by Chemical Treatment of Ti Substrate with a H_2O_2/HCl Solution, *Biomaterials*, **23**, 1353-1357 (2002).

[4]A. Osaka, K. Tsuru, and S. Hayakawa, Titania derived from combined chemical and thermal treatments of titanium: in vitro apatite-forming ability, *Phosph. Res. Bull.*, **17**, 130-141 (2004).

[5]A. Sugino, K. Uetsuki, K. Tsuru, S. Hayakawa, A. Osaka, Surface Topography Designed to Provide Osteoconductivity to Titanium after Thermal Oxidation, *Mater. Trans.*, **49**, 428-434 (2008).

[6]A. Sugino, C. Ohtsuki, K. Tsuru, S. Hayakawa, T. Nakano, Y. Okazaki, A. Osaka, Effect of spatial design and thermal oxidation on apatite formation on Ti-15Zr-4Ta-4Nb alloy, *Acta Biomaterialia*, **5**, 298-304 (2009).

[7]A. Sugino, K. Tsuru, S. Hayakawa, K. Kikuta, G. Kawachi, A. Osaka and C. Ohtsuki, Induced deposition of bone-like hydroxyapatite on thermally oxidized titanium substrates using a spatial gap in a solution that mimics a body fluid, *J. Ceram. Soc. Japan*, **117**, 515-520 (2009).

[8]K. Anselme, Osteoblast adhesion on biomaterials, *Biomaterials*, **21**, 667–681 (2002).

[9]J. O'Brien, I. Wilson, T. Orton, and F. Pognan, Investigation of the Alamar Blue (resazurin) fluorescent dye for the assessment of mammalian cell cytotoxicity, *Eur. J. Biochem.* **267**, 5421-5426 (2000).

[10]KT. Bowers, JC. Keller, BA. Randolph, DG. Wick, and CM. Michaels., Optimization of surface micromorphology for enhanced osteoblast responses in vitro, *Int J Oral Maxillofac Implants*, **7**, 302–310 (1992).

[11]JY. Martin, Z. Schwartz, TW. Hummert, DM. Schraub, J. Simpson, J. Lankford Jr, DD. Dean, DL. Cochran, and BD. Boyan. Effect of titanium surface roughness on proliferation, differentiation, and protein synthesis of human osteoblast-like cells (MG63), *J Biomed Mater Res*, **29**, 389–401 (1995).

[12]X. Wen, X. Wang, N. Zhang, Microrough surface of metallic biomaterials: A literature review. *Biomed Mater Eng*, **6**, 173-189 (1995).

[13]A. Curtis, C. Wilkinson, Review: Topographical control of cells, *Biomaterials*, **18**, 1573-1583 (1997).

[14]DD. Deligianni, N. Katsala, S. Ladas, D. Sotiropoulou, J. Amedee, and YF. Missirlis, Effect of surface roughness of the titanium alloy Ti–6A1–4V on human bone marrow cell response and on protein adsorption, *Biomaterials*, **22**, 1241–1251 (2001).

[15]TJ. Webster, C. Ergun, RH. Doremus, RW. Siegel, and R. Bizios, Specific proteins mediate enhanced osteoblast adhesion on nanophase ceramics, *J Biomed Mater Res*, **51**, 475-483 (2000).

FREEFORM FABRICATION AND STRUCTURAL CONTROLS OF ALUMINA DENTAL-CROWN MODELS BY USING STEREOLITHOGRAPHY

Mitsuyori Suwa[a], Soshu Kirihara[a], and Taiji Sohmura[b]
a) Joining and Welding Research Institute, Osaka University, Ibaraki, Osaka 567-0047, Japan
b) School of Dentistry, Osaka University, Suita, Osaka 565-0871, Japan

ABSTRACT
    Three-dimensional dental-crown models composed of alumina and glass composite were fabricated successfully by using stereolithography. Photo sensitive acrylic resins including alumina particles of 0.17 and 1.8μm in diameters at 40 and 70 volume percent were spread on a metal substrate with 30 and 70μm in layer thickness through mechanical knife edge moving, respectively. These precursors were dewaxed at 600 °C and sintered at 1500 and 1700 °C in the air. Moreover, alumina bending test specimens of 1.2×4×20 mm in dimensions were fabricated by using the similar materials and processes. The average bending strength was about 216 MPa. Furthermore, La_2O_3-B_2O_3-Al_2O_3-SiO_2 glasses were coated on the alumina specimens to close micro cracks on the surface. The ceramic bulks with the glass powder spreading were heated at 1100 °C for 2hs in the air. The glass coated samples were sand blasted. The glass coated alumina bulks showed the higher mechanical property of 415 MPa in average bending strength.

INTRODUCTION
    In recent years, all ceramic dental-crowns have been focused in dental restoration areas for dentures and crowns, because of superiorities on desirable aesthetics and bio- compatibility compared with traditional metal-ceramic restorations. A huge variety of methods have been considered to produce ceramic structures[1-3]. Especially, cutting works are currently prevailed as the method of ceramic crowns forming. However, the cutting tools exhausted easily because of high hardness of the ceramic materials. In addition, only one single dental-crown with the complicated shape adjusted for individual persons can be formed through one operation of the cutting methods. In contrast with this, rapid prototyping processes of layer lamination methods are possible to create different components with various shapes simultaneously. Our research group had succeeded in fabrication of three dimensional ceramic crown models by stereolithography[4]. A dense Al_2O_3 dental-crown model has been obtained by heat treatment for the green body composed of nanometer sized Al_2O_3 particles dispersed in acrylic resin fabricated by using the stereolithography. However, the measured flexural strength of the sintered crown model has been insufficient for clinical practice. In this study, the Al_2O_3 particle size effects for the flexural strength of the sintered bulks and the reproducibility of the complex dental-crown model were investigated.

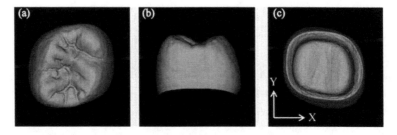

Figure 1   A computer graphic model of a dental-crown obtained from a CT scanning system. (a), (b) and (c) are top, side and bottom views, respectively.

EXPERIMENTAL PROCEDURE

Figure 1 show the graphical views of the dental-crown of an upper molar tooth for different directions obtained by a CT scanning systems. The three dimensional graphic model was converted into the stereolithography format of STL file, and sliced into a series of two dimensional data with uniform thickness by using the CAD application as illustrated in Figure 2-(a). These numerical data were transferred into the stereolithography equipment (SCS-300P, D-MEC, Japan,). Figure 2-(b) show the schematically illustrated fundamental processes in the stereolithography. The photo sensitive acrylic resins including Al_2O_3 particles of 0.17 and 1.8 μm in diameters at 40 and 70 vol. % were spread on a substrate with 30 and 70 μm in layer thickness through a mechanical knife edge, respectively. An ultraviolet laser beam of 355 nm in wavelength was scanned on the slurry surface according to the cross sectional data. The laser beam was adjusted 100 μm in spot diameter and 100 mW in power. The photo sensitive resin slurries could be polymerized with including the ceramic particles. By the layer stacking processes, the solid object of the dental-crowns were fabricated. The composite precursors were dewaxed at 600 °C for 2hs with the heating rate 0.5 °C/min, and sintered at 1700 °C for 2hs with 8 °C/min in the air atmosphere. Figure 3 shows a typical heat treatment pattern. The densities were measured by Archimedes method. The sintered sample was heated in the La_2O_3-B_2O_3-Al_2O_3-SiO_2 glass ceramic powder (In-Ceram-Alumina, Zahnfabrik, Germany) at 1100 °C for 2hs in the air to infiltrate the molten phase into the micro cracks or pores by osmotic pressures. This glass powder is widely used for the ceramic coating on metal teeth in the dentistry. The excess glass powder on the surface was eliminated by sand blasting. Subsequently, the bulk samples of 1.2×4×20 mm were formed to measure the flexural strength. The acrylic bulks including with Al_2O_3 particles of 1.8 μm in diameter at 70 vol. % were sintered at 1400, 1500, 1600 and 1700 °C. The microstructures of sintered alumina were observed by scanning electron microscope (SEM). The flexural strength was measured for the glass coated samples, and the fracture surfaces were observed by SEM.

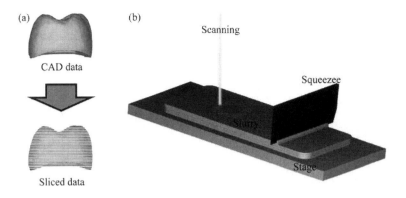

Figure 2   Schematic illustrations of a data slicing (a) and a stereolithography process (b) .

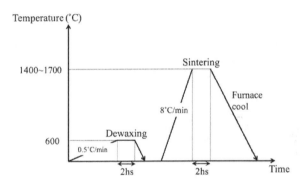

Figure 3   Heat treatment patterns for acrylic resin precursors with Al_2O_3 particles dispersion.

RESULTS AND DISCUSSION

Figure 4-(a), (b) and (c) show the different direction views of the dental-crown precursor including with the Al_2O_3 particles of 0.17 μm in diameter at 40 vol. % fabricated by the stereolithography. The average dimension tolerance was within approximately 100 μm. The grooves at the top surface of green body have been precisely shaped, and the peripheral edge on the bottom formed sharply. Large stacking faults of layers were not observed on the surface. Figure 4-(d), (e) and (f) show the sintered body shapes of the dental-crown model along the different directions. The measured relative density reached approximately 98 %. X-ray diffraction peaks of carbon or carbides were not observed for the sintered samples. Nano-size Al2O3 powders were considered to be sintered effectively at 1500 °C. The linear shrinkage ratios of the horizontal and vertical axes resulted in approximately 24 and 28 %,

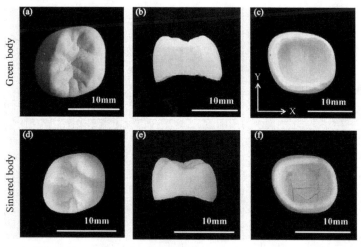

Figure 4 External appearances for different directions of an acrylic green body including with the Al₂O₃ particles of 0.17 μm in diameter (a)-top, (b)-bottom, (c)-side and a sintered body (d)-top, (e)-bottom, (f)-side.

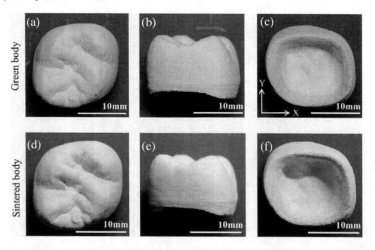

Figure 5 The appearances for different directions of the acrylic green body including with the Al₂O₃ particles of 1.8 μm in diameter (a)-top, (b) -side, (c) -bottom and the sintered body (d)-top, (e) -side, (f) -bottom.

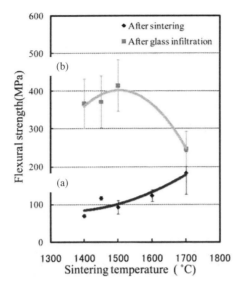

Figure 6 The relationship between the sintering temperatures and flexural strengths of the Al_2O_3 bulks. (a) After sintering (b) After glass infiltration

respectively. The large cracks or pores were not observed on the top and side smooth surfaces. However, some cracks were formed on the inside surface. The distortion in the shrinkage during the heat treatment is considered to cause the cracking. The average flexural strengths before and after the glass infiltration were 64 ± 11 and 197 ± 24 MPa. These are lower level comparing with the required values of 400 MPa for the artificial tooth. Figure 5-(a), (b) and (c) show the green body shapes of the dental-crown model composed of photo sensitive resin including with the Al_2O_3 particles of 1.8µm in diameter along different directions. The average dimension tolerance was within approximately 100 µm. Figure 5-(d), (e) and (f) show the sintered body shapes. The relative density reached approximately 97 %. The X-ray diffraction peaks of carbon and carbides were not identified. The linear shrinkage ratios of horizontal and vertical axes were approximately 7 and 9 %, respectively. The larger shrinkage for the vertical axis is considered to be occurred by the gravity effect. Linear shrinkage ratios were decreased comparing with nano-particles sintering as shown in Fig. 4, because the alumina particle were dispersed into the acrylic resin at 70 % of higher volume fraction. The ceramics particle dispersion with the high volume fraction realized restraints of the shape deformations during the dewaxing and sintering.

Figure 6-(a) shows the measured results of the flexural tests for the sintered alumina bulks. The flexural strengths of as-sintered bulks were increased with the elevation of sintering temperatures.

The maximum value of 216±26 MPa was obtained for the sintered sample at 1700 °C. In contrast with this, the flexural strengths of glass infiltrated alumina

Figure 7   SEM micrographs of the Al$_2$O$_3$ bulk surfaces sintered at (a)-1500, (b)-1600 and (c)-1700 °C for 2hs in the air atmosphere.

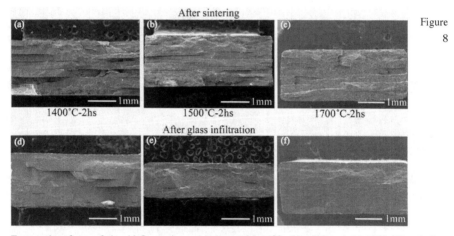

Figure 8

Fractured surfaces of the Al$_2$O$_3$ as-sintered bulks at (a)-1400, (b)-1500 and (c)-1700 °C and glass infiltrated bulks (d), (e) and (f) for (a), (b) and (c), respectively.

bulks show the maximum value of 415±68 MPa at 1500 °C in sintering temperature as shown in Figure 6-(b). Figure 7 shows the surface microstructures of the sintered alumina. The Al$_2$O$_3$ crystal grains grew prominently at the higher sintering temperatures. The grain growth is considered to fill the defect pores in the microstructure. Figure 8 show the microstructures of the fractured surfaces in the Al$_2$O$_3$ bulks with or without the glass infiltration. The parallel macro crack formations are more prominent in the as-sintered samples at the lower temperatures. Comparing with this, the cracks are obscured in the glass infiltrated bulks. As shown in Figure 7-(a), the sintered Al$_2$O$_3$ structure at 1500

°C has many open pores. Therefore, the molten glass material is considered to infiltrate into the porous structure smoothly, and form the network structure increasing the flexural strength.

CONCLUSION

A three-dimensional dental-crown model composed of 70 vol. % Al_2O_3 particles dispersed acrylic resin was successfully fabricated by using stereolithography. After dewaxing at 600 °C and sintering at 1700 °C, a dense Al_2O_3 ceramic body was obtained without deformation. The relative density was approximately 97%. The maximum flexural strength of the sintered specimens was 216±26 MPa. The mechanical property could be increased at 415±68 MPa by a glass infiltration technique.

ACKNOWLEDGMENTS

This study was supported by Priority Assistance for the Formation of Worldwide Renowned Centers of Research - The Global COE Program (Project: Center of Excellence for Advanced Structural and Functional Materials Design) from the Ministry of Education, Culture, Sports, Science and Technology (MEXT), Japan.

REFERENCES

[1]J. Stampfl, H. C. Liu, S. W. Nam, K. Sakamoto, H. Tsuru, S. Kang, A. G. Cooper, A. Nikel, F. B. Prinz, Rapid prototyping and manufacturing by gelcasting of metallic and ceramic slurries, *Materials Science and Engineering*, **A334**, 187-192 (2002).

[2]H. H. Tnag, F. H. Liu, Ceramic laser gelling, *Journal of the European Ceramic Society*, **25**, 627-632 (2005).

[3]R. Zauner, Micro powder infection moulding, *Microelectric Engineering*, **83**, 1442-1444 (2006).

[4]M. Ishikawa, S. Kirihara, Y. Miyamoto, T. Sohmura, Freeform fabrication of alumina dental-crown model by using stereolithography, *Advanced Processing and manufacturing tecnnologies for Structural and Multifunctional Materials II, Ceramic Engineering and Science Proceeding* **Volume 29**, Issue 9 (2008).

MOISTURE EFFECT ON TUBULAR ALUMINA TOUGHENED ZIRCONIA CERAMIC FOR IMPLANT CASING

Wai Man Grace So and Guangqiang Jiang
The Alfred E. Mann Foundation for Scientific Research
Santa Clarita, California, USA

ABSTRACT
Alumina toughened zirconia (ATZ, 80 wt. % 3YTZP [3 mol% yttria stabilized tetragonal zirconia polycrystals] with 20 wt. % alumina) tubular hermetic ceramic cases were built and tested with simulated accelerated aging tests in pressurized steam. The objectives were to characterize the strength change in relation to aging time and to eventually predict and improve their lifetime in vivo. Monoclinic phase transformation was observed on the outer exposed surface with a positive linear monoclinic layer thickness relationship to aging time. This linear relationship also resulted in increasing outer diameter and decreasing flexural fracture load with aging time. Apart from the well-known monoclinic content, another detrimental trait was identified—inclusions. The flexural fracture loads, measured by "four-line bending", were lower in ceramic cases having inclusions on the fracture surface. The fracture initiation inclusion is typically found directly under the monoclinic transformation interface within the inner loading points. Depending on the type and size of inclusions, some could lower the fracture load below a critical value. A challenge in screening out such tubes is that the inclusions are embedded in the ceramic bulk. Suggestions include X-ray CT-scanning and sampling to estimate the inclusion status in a batch of ceramic tubes.

INTRODUCTION
    ATZ ceramic was chosen to encase the electronic package of an implantable medical device because of its excellent mechanical properties and its radio frequency transparency. The material has demonstrated the combined advantages of the hardness and wear resistance of alumina as well as the high biaxial bending strength and flexural toughness of zirconia, which is often used in joint replacements[1]. However zirconia also experiences low temperature degradation (LTD) in response to stress and/or moisture that leads to well known destructive tetragonal-to-monoclinic phase change. Furthermore, the phase change leads to volume expansion and consequently to stresses and fracture. The volume expansion due to phase transformation for 3YTZP is approximately 4%[2]. ATZ is not yet proven experimentally to be acceptable for implant applications. To test, study, and characterize the aging behavior of the implant design in vivo, simulated accelerated aging tests in the form of soaking tests were carried out on a total of more than 80 casing samples.
    ATZ ceramic tubes were attached to metal caps on both ends by brazing and welding for implant casing (Figure 1). The implant acts as an electro-neuro-stimulator and will be inserted and positioned in muscle or excitable tissue. This implant is not intended to bear load and wear as much as zirconia joint replacements, but will be exposed to static and dynamic loading within the human body during daily activities.

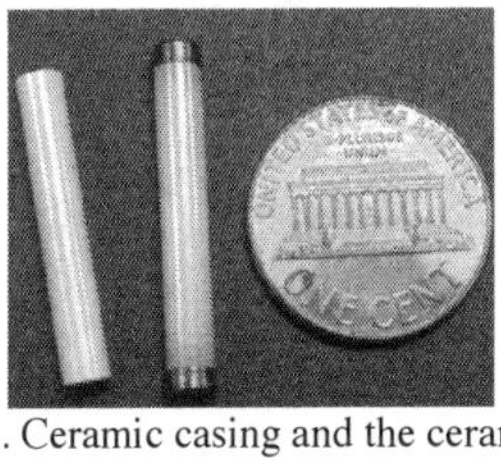

Figure 1. Ceramic casing and the ceramic tube

The response of the implant towards in vitro aging is characterized by monitoring the change in monoclinic content and the accompanying volume expansion. The increasing monoclinic content was measured by X-ray diffraction (XRD) between aging test intervals. The final transformed monoclinic layer thickness was measured by scanning electron microscopy (SEM) on the fracture surfaces after the samples were failed in four-line bending. The outer diameter of the casings was measured to determine the volumetric expansion. To assure that the volumetric expansion of the ceramic tube is within allowable limits for the metal caps on both ends, helium leak testing was performed to verify that the bond was still hermetic and that the ceramic to metal joints are intact. When each sample finished the final soaking interval, four-line bending to fracture was performed as a quantitative indication of its mechanical properties. Visual inspection using optical microscope was also performed before, during, and at the conclusion of the tests to check for cracks and any other defects on the samples to find inclusions and scratches. Energy-dispersive X-ray spectroscopy (EDS) was performed to determine the elemental composition on the fracture surface of the defects.

EXPERIMENTAL METHODS

The simulated accelerated aging tests were performed in deionized water vapor using a Yamato autoclave at 127°C. A 171-hour exposure to these conditions is equivalent to 10-years in vivo aging based on the Arrhenius factor reported previously on 3YTZP materials[3]. This Arrhenius factor is a conservative estimate, based on the activation energy value obtained for an ATZ ceramic reported by Schneider[4].

A group of ten samples were aged 171 hours to a 10-year equivalent in vivo exposure. Another group of ten samples were aged to a 20-year in vivo equivalent. Eight groups of samples, each containing ten casings, were aged 10 to 80 years equivalent. Measurements and visual inspections were performed before and after each aging interval. One additional group of three samples was aged at a shorter interval (25-hours each), ranging to approximately 25-year in vivo equivalent, to monitor the monoclinic transformation rate at a shorter interval. Two other groups of unaged ceramic cases and ceramic tubes were used as control samples.

The surface monoclinic content was measured by XRD using Siemens D500, Cu K-alpha radiation. Volume expansion was verified by measuring the outer diameter of the post-exposure ceramic, both at the middle and near the metal caps, of all the samples. Optical microscopy up to 90 times magnification was used in visual inspection. Four-line bending (Figure 2) using metal bands as load lines was employed instead of four-point bending to avoid stress concentration at the load points. FEI Inspect S SEM and Thermo Fisher Scientific UltraDry EDS were employed in this study.

Figure 2. Four-line bending apparatus using metal bands as loading lines

RESULTS AND DISCUSSIONS

The surface monoclinic transformation rate is presented in Figure 3. All the samples initially have 0% monoclinic phase that increases during aging. The maximum monoclinic content measured post-exposure was 70% on the surface.

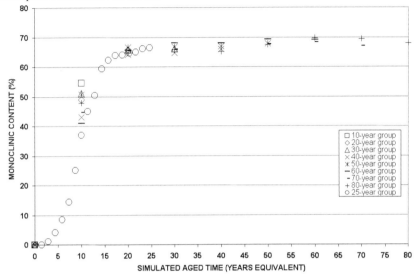

Figure 3. Monoclinic phase content measured from the outer ceramic surface

Although saturation on the surface (approximately 67% of monoclinic content) was reached at approximately 20-years equivalent exposure, the monoclinic phase transformation continues in the ceramic. The thickness of the monoclinic layer continues to increase with exposure time (Figure 4). The thickness of the monoclinic layer increases approximately 1 micron per year.

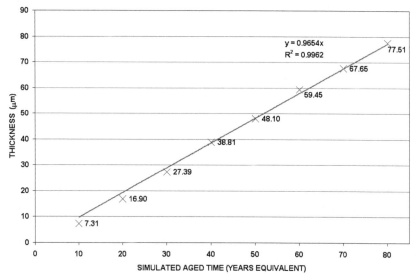

Figure 4. Post-exposure monoclinic phase thickness increases approximately linearly

The change in outer diameter of the ceramic casing of both middle portion and the end portion near the metal caps are shown in Figure 5. Each data point was averaged from 30 readings, 3 readings each from 10 samples in each group. The initial dimensions of tubes prior to assembly are also presented as a reference. The outer diameter at the middle portion is slightly larger than the diameter at the end, rendering a slight "barrel" shape. Both outer diameters increased in response to aging. If the trend was to be taken as linear, the R-squared values will be 0.99 and 0.98 for middle and end outer diameters respectively. This diametral growth rate is related to this particular configuration of ceramic tube. As a result of this diametral expansion, an average circumferential strain of 0.85% was experienced near the tubes ends in the 80th year-equivalent. This strain will create stress on the metal caps.

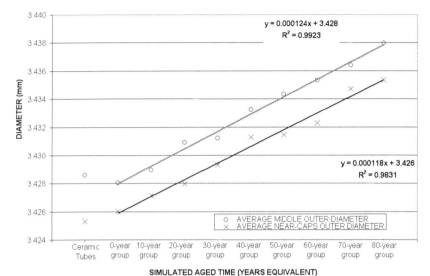

Figure 5. Diametral change of tube post-aging

After the samples were aged, they were loaded in four-line bending to failure. The failure load (Figure 6) exhibits a decreasing trend with increased aging time. The failure loads for assembled casings are lower than as-supplied ceramic tubes. The variation within each aging group decreases with the aging time. If the decreasing trend was to be taken as linear, the R-squared value is 0.88. Comparing the 0th and 80th year equivalent average fracture loads, the ceramic casing retained 67% of its original strength after 80-year equivalent aging. This decrease in ceramic casing strength, due to aging, should be accounted for in the design phase. All but one sample initiated fractures at the middle portion of top surfaces between the inner metal bands (Figure 2) in longitudinal direction. This is where the maximum tensile stress is created by four-line bending. The exceptional sample (lowest load in Group 20-year) is shown later in Figure 8 where a strange inclusion initiated the premature fracture.

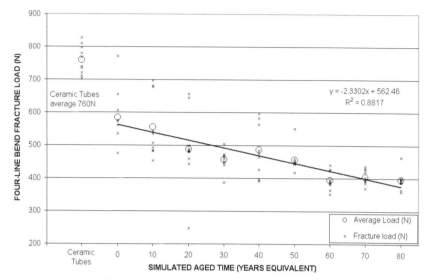

Figure 6. Strength of aged casings decreases with time

SEM and EDS analyses were performed on the fracture surfaces of the samples with the lowest failure strength in each group. Inclusions were found on some of the fracture surfaces at the tensile load side, but not on all fracture surfaces. The majority of the inclusions is sodium-rich and is located just below the tetragonal-to-monoclinic phase transformation interface (Figure 7). However, the sample with the lowest fracture load (248N) in the 20-year group contained a rare inclusion sized about 20 by 60 microns (Figure 8). It has the same composition as the surrounding ATZ bulk sample and is not located under the phase transformation interface.

Figure 7. Inclusion containing sodium found underneath the tetragonal-to-monoclinic transformation interface

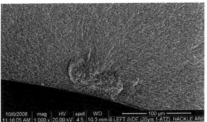

Figure 8. Inclusion having identical chemical compositions as ATZ found on the fracture surface of the weakest sample

Attempts have been made to screen such inclusions. X-Ray CT scan (Figure 9) is effective when the inclusion's density is significantly different from ATZ bulk ceramic. Another suggestion is sampling incoming as-received ceramic tubes and performing destructive tests to check for inclusions.

Figure 9. X-Ray CT-scan showing an embedded inclusion in a ceramic tube

CONCLUSION

The tubular ATZ ceramic casings degraded in moist environment through tetragonal-to-monoclinic phase transformation. This leads to several measurable changes. Firstly, the monoclinic phase thickness increases linearly, even after the surface monoclinic content saturation was reached. Secondly, the outer diameters of the ceramic casing increased due to the volume expansion. Thirdly, the flexural strength, as measured by four-line bending, decreases with aging time. In addition to the above observable degradation, hidden defects also reduce the strength of the casing: embedded inclusions. A combination of X-ray CT scan screening and accelerated aging should be part of the process to ensure that the ceramic casing can tolerate the degradation in vivo. Further investigation is yet required to determine the impact of inclusions size on ceramic casing strength.

ACKNOWLEDGEMENTS

The authors wish to thank Dr. G. Schnittgrund, P. Teich, B. Dearden, A. Antalfy, J. Gutierrez and S. O'Grady for their assistance with the study.

REFERENCES

[1]Sabine Begand, Thomas Oberbach, and Wilfried Glien, ATZ – A New Material with a High Potential in Joint Replacement, *Key Engineering Materials*, **Vols. 284-286,** 983-6 (2005)

[2] S. Deville, L. Gremilllard, J. Chevalier, and G. Fantozzi, A critical Comparison of Methods for the Determination of the Aging Sensitivity in Biomedical Grade Yttria-Stabilized Zirconia, *Biomed. Mater. Res. Part B*, **72B,** 139-45 (1999)

[3]G. Jiang, K. Purnell, J. Schulman and P. Mobley, Accelerated Life Tests and In-Vivo Test of 3Y-TZP Ceramics, *Proceedings of the Materials & Processes for Medical Devices Conference 2003*, 477-82 (2003)

[4]Jens Schneider, Sabine Begand, Ralf Kriegel, Christian Kaps, Wilfried Glien, and Thomas Oberbach, Low-Temperature Aging Behavior of Alumina-Toughened Zirconia, *J. Am Ceram. Soc.*, **91 [11],** 3613-8 (2008)

Author Index